# Active Polymers

T0331194

MATERIALS RESEARCH SOCIETY
SYMPOSIUM PROCEEDINGS VOLUME 1190

# Active Polymers

Symposium held April 14–17, 2009, San Francisco, California, U.S.A.

**EDITORS:**

## Andreas Lendlein

GKSS Research Centre Geesthacht GmbH
Teltow, Germany

## V. Prasad Shastri

Albert-Ludwigs University of Freiburg
Freiburg, Germany

## Ken Gall

Georgia Institute of Technology
Atlanta, Georgia, U.S.A.

Materials Research Society
Warrendale, Pennsylvania

# CAMBRIDGE
UNIVERSITY PRESS

University Printing House, Cambridge CB2 8BS, United Kingdom

One Liberty Plaza, 20th Floor, New York, NY 10006, USA

477 Williamstown Road, Port Melbourne, VIC 3207, Australia

314-321, 3rd Floor, Plot 3, Splendor Forum, Jasola District Centre, New Delhi - 110025, India

79 Anson Road, #06-04/06, Singapore 079906

Cambridge University Press is part of the University of Cambridge.

It furthers the University's mission by disseminating knowledge in the pursuit of education, learning and research at the highest international levels of excellence.

www.cambridge.org
Information on this title: www.cambridge.org/9781605111636

Materials Research Society
506 Keystone Drive, Warrendale, PA 15086
http://www.mrs.org

© Materials Research Society 2009

First published 2009
First paperback edition 2012

Single article reprints from this publication are available through University Microfilms Inc., 300 North Zeeb Road, Ann Arbor, MI 48106

CODEN: MRSPDH

*A catalogue record for this publication is available from the British Library*

ISBN 978-1-605-11163-6 Hardback
ISBN 978-1-107-40816-6 Paperback

## SHAPE-MEMORY POLYMERS

*Invited Paper

## STIMULI-SENSITIVE HYDROGELS

## BIOMATERIALS

*Invited Paper

## STIMULI-SENSITIVE SYSTEMS, ACTUATORS, AND SENSORS

*Invited Paper

# PREFACE

The field of polymer research is progressing rapidly from passive materials providing a certain set of properties to active polymers, which provide, receive, and respond to signals from their environment. This includes interactions with molecules, biological systems, and physical stimuli. Research in active polymers has been driven by the increasing demand for intelligent materials, especially in biomedical and aerospace applications. Tremendous progress in synthesis, analytics, and molecular modeling enables scientists today to develop active polymer systems in a knowledge-based approach.

Biological systems might serve as blue prints for biomimetic and bionic solutions. Interdisciplinary approaches combining the expertise of chemists, physicists, biologists, pharmacologists, and materials engineers lead to innovative material concepts, e.g., enabling morphing structures or "living" textiles. Emerging active polymers respond to a range of stimuli from changes in pH and temperature to light and magnetic fields. Interactive behavior can also be stimulated by small bioactive molecules or influences of complex biological systems. Remote and on-demand control is envisioned, and model systems have been developed.

Symposium NN, "Active Polymers," held April 14–17 at the 2009 MRS Spring Meeting in San Francisco, California, was the first MRS symposium focused on active polymers. A highly interdisciplinary scientific community used this great opportunity to gather and discuss the topics: shape-memory polymers; shape-changing polymers; responsive hydrogels; stimuli-sensitive systems; intelligent polymers in biological systems; polymer-based actuators, sensors, and switches; active surfaces; and biomedical applications of active materials, especially for tissue regeneration and controlled drug release. Symposium NN "Active Polymers" far exceeded our expectations in participation, diversity of topics, and duration, with 60 oral presentations and 43 posters spanning twelve sessions.

Notable presentations included shape-memory nanocomposites by Richard A. Vaia, and color-changing shape-memory polymers by Patrick T. Mather. Duncan Maitland reported on design and realization of biomedical devices based on shape-memory polymers. The application of shape-memory polymers as space deployables and novel actuators was presented by Steven C. Arzberger. Rein V. Ulijn presented about exploiting enzymes in responsive materials and nanofabrications.

We kindly acknowledge financial support for this symposium from

Fluoron GmbH, a company of the Geuder® Group
Fresenius Medical Care Deutschland GmbH
GKSS-Forschungszentrum Geesthacht GmbH
Invitek GmbH, Biotechnology & Biodesign Ltd
Whatman Ltd, Life Sciences - GE Health Care

Andreas Lendlein
V. Prasad Shastri
Ken Gall

October 2009

# MATERIALS RESEARCH SOCIETY SYMPOSIUM PROCEEDINGS

# MATERIALS RESEARCH SOCIETY SYMPOSIUM PROCEEDINGS

**Prior Materials Research Society Symposium Proceedings available by contacting Materials Research Society**

# Shape-Memory Polymers

Mater. Res. Soc. Symp. Proc. Vol. 1190 © 2009 Materials Research Society          1190-NN01-06

## Multiphase Polymer Networks With Shape-Memory

Steffen Kelch[1], Marc Behl[2], Stefan Kamlage[2] and Andreas Lendlein[2]
[1]Sika Technology AG, Tüffenwies 16, CH-8048 Zürich, Switzerland
[2]Center for Biomaterial Development, Institute of Polymer Research, GKSS Research Center Geesthacht GmbH, Kantstr. 55, 14513 Teltow, Germany

### ABSTRACT

Tailoring of properties and functions of shape-memory polymer networks to the requirements of specific applications demands a knowledge-based approach. A comprehensive database enabling the analysis of structure-property relationship is obtained by the systematic variation of molecular parameters. Here, we investigated the influence of the nature of thermal transition on the shape-memory behavior of polymer networks. Furthermore, additional amorphous phases were introduced enabling tailoring of elastic properties especially in the temporary shape. The structure property relationships were derived for different designs of such multiphase polymer network architectures.

### INTRODUCTION

A recent development in shape-memory polymer technology is the design of shape-memory polymers as multifunctional materials. While the intrinsic properties of a material are determined by its molecular structure, the functionality of material arises from the combination of a suitable polymer architecture and a specific process [1]. Multifunctionality is the unexpected combination of two or more functions of a material. While usually functionalities are combined on purpose, the term unexpected results from the fact that such combinations are not apparent. An interesting example of such multifunctional materials are hydrolytically degradable shape-memory polymers, which combine the functionalities degradability and the shape-memory capability. While degradability could be obtained by the introduction of easily cleavable bonds, the shape-memory capability requires a suitable polymer network architecture in combination with an appropriate programming procedure. Several previous articles described the programming process and the characterization of the shape-memory effect and gave definitions for shape fixity ($R_f$) and shape recovery rates ($R_r$) [2-6]. In this article, we focused on the design of suitable polymer networks architectures enabling the shape-memory capability and the parameters to control the shape-memory properties. For a better comparability of the results presented, we will limit our considerations to polymer networks based on macrodimethacrylates.

### DISCUSSION

#### Polymer network architectures

Suitable polymer networks structures having shape-memory capability are displayed in Scheme 1. General parameters for controlling the shape-memory behaviour of covalently crosslinked polymer networks are the functionality of the crosslinks, the network chain segment length, the nature of the switching segment and the number of phases. The functionality of crosslinks and the chain segment length control the crosslink density and influence in this way the mechanical properties while the nature of the switching segment influences the hydrolytic degradation as well as the characteristics of the shape-memory effect. The switching domains can be either related to a melting transition ($T_m$) or a glass

3

transition ($T_g$) resulting in either a sharp melting transition ($T_m$) or a glass transition ($T_g$) extending over a broader temperature interval.

| polymer network class | crystallizable switching segment | amorphous switching segment |
|---|---|---|
| 1 | **(co)polymer network**<br>poly($\varepsilon$-caprolactone) PCL<br>poly($\varepsilon$-caprolactone-co-glycolide) PCG  | **(co)polymer network**<br>poly[($L$-lactide)-$ran$-glycolide] PLLG  |
| 2 | **AB copolymer network**<br>PCL/acrylate<br>PCG/acrylate  | **AB copolymer network**<br>PLLG/acrylate  |
| 3 | | **ABA block copolymer network**<br>poly[($rac$-lactide)-$block$-PPO-$block$-($rac$-lactide)]  |

Scheme 1: (Multiphase) polymer network architectures and suitable starting materials

In the following we will focus on telechelics having two reactive groups as starting materials. In polymer networks with a crystallizable switching segment the molecular length can be used for controlling the switching temperature as well. Appropriate macromonomers for polymer networks of class 1 are listed in Scheme 1.

Another strategy for influencing the mechanical properties of the shape-memory polymer networks was the incorporation of a second comonomer which leaded to the formation of an additional amorphous phase, which contributed to the overall elasticity of the polymer network (class 2). Such polymer networks were prepared from crystallizable as well as amorphous switching segments. In such polymer networks the phases determining the shape-memory capability as well as the phases providing the additional elasticity were on the molecular level distributed among different monomers. In amorphous polymer networks from poly[($rac$-lactide)-$block$-polypropyleneglycol-$block$-($rac$-lactide)]dimethacrylate it was demonstrated, that the phase providing additional elasticity could both be provided by the same linear macromonomer (class 3).

**Synthesis**

Polymer network synthesis could be realized by covalently crosslinking of macrodimethacrylates. The polymerization could be initiated by a thermal radical initiator or by UV light. Photopolymerization is favorable with respect to a biomedical application of the shape-memory materials as harmful effects of the initiator as well as from its degradation products are avoided [7]. Both kinds of polymerizations required the formation of macrodimethacrylates from macrodiols. Such macrodiols were prepared in a ring-opening polymerization (ROP) using hydroxytelechelic initiators. Copolymerization enabled control over the nature of the switching segment. While copolymerization of $\varepsilon$-caprolactone with diglycolide resulted in semicrystalline macrodiols, amorphous macrodiols were obtained from the copolymerization of L,L-dilactide and diglycolide especially when a transesterification catalyst was added. The different sets of comonomers used in the synthesis of macrodiols are shown in Scheme 2.

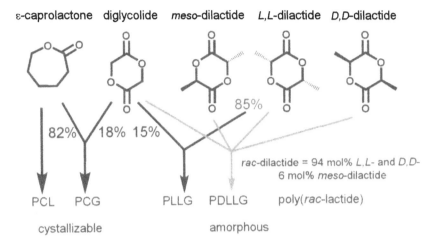

Scheme 2: Synthesis macrodiols from lactones and cyclic diesters and influence of starting materials on nature of thermal transition associated to switching domains.

The nature of the thermal transition related to the phase formed by the macrodiols could be additionally controlled by using a catalyst during ROP. The copolymerization of ε-caprolactone and diglycolide without addition of a transesterification catalyst resulted in crystallizable macrodiols having a blocky sequence structure. An amorphous macrodiol with a random copolymer structure was obtained, when the same copolymerization is performed in the presence of a catalyst capable of transesterification.

Macrodimethacrylates were obtained by esterification of the synthesized macrodiols with methacryloyl chloride. Depending on the macrodiols these macrodimethacrylates can be used for polymer network formation with amorphous or crystalline switching domains. The polymer networks according to type 1 with $T_{trans} = T_m$ were formed from acrylate or methacrylate functionalized poly(ε-caprolactone)diols (PCL) or poly[(ε-caprolactone)-co-glycolide]diols (PCG) while polymer networks with $T_{trans} = T_g$ were formed from poly[(L-lactide)-ran-glycolide]diols (PLGA). In both cases polymerization was initiated by irradiation with UV-light.

AB copolymer networks having a second amorphous phase providing additional elasticity according to type 2 were prepared in case of $T_{trans} = T_m$ from oligo[(ε-caprolactone)-co-glycolide]dimethacrylate and n-butyl acrylate. n-Butyl acrylate was chosen as comonomer as it provides an additional amorphous and non-crystallizable phase of poly(n-butyl acrylate) associated to a low glass transition temperature [$T_g$ of poly(n-butyl acrylate) = -55 °C]. Copolymerization of poly[(L-lactide)-ran-glycolide]dimethacrylate with n-ethyl acrylate, n-butyl acrylate or n-hexyl acrylate yielded AB copolymer networks $T_{trans} = T_g$. Both polymer network systems having $T_{trans} = T_m$ or $T_{trans} = T_g$ were obtained by photopolymerization.

ABA networks with $T_{trans} = T_g$, in which the elasticity providing phase is incorporated in the polymer main chain of the macromonomer could be formed by suitable oligomers of different molar mass. ROP of rac-dilactide with poly(propylene oxide) as macroinitiator in presence of dibutyltin oxide as catalyst was applied for macrodiol formation. The macrodimethacrylates were obtained by endgroup functionalization of these ABA triblock copolymer diols having a middle block of poly(propylene oxide) with methacrylate groups. The final polymer networks based on such macrodimethacrylates were generated by photocrosslinking using UV-light.

## Thermal Properties

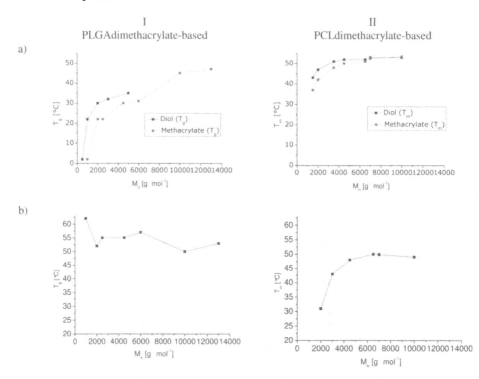

Figure 1: Thermal transition temperatures of macrodiols, macrodimethacrylates (a) and polymer networks (b) having amorphous (I) and crystallizable switching phases (II)

Macrodiols and macrodimethacrylates of PLGA and PCL exhibited a similar correlation of transition temperature with molecular weight of the oligomers. The telechelics reached individual maximum $T_{trans}$ with increasing molecular weight ($M_n$) (Figure 1). A similar behavior was determined for $T_m$ of PCL networks which increased with increasing molecular weight of the oligomers used as starting material. In case of PCG the dependence of $T_m$ for macrodimethacrylates and polymer networks displayed a similar correlation between $T_m$ and $M_n$. In contrast, the glass transition temperatures of PLGA networks were independent from the $M_n$ of the precursor if the molecular weights of the macrodimethacrylate precursors are above 2 kD.

In AB-type shape-memory polymer networks having a crystallizable switching domains $T_m$ is decreasing with increasing $n$-butyl acrylate content. Increasing $n$-butyl acrylate content caused a decrease in crystallinity and in melting enthalpy $\Delta H_m$ of the polymer networks. The content of acrylate also influences the mechanical properties of the polymer networks. These polymer networks became more elastic with increasing amount of *n*-butyl acrylate. Such a behavior was shown for AB networks from *n*-butyl acrylate and poly(ε-caprolactone)dimethacrylate or poly[(ε-caprolactone)-*co*-glycolide]dimethacrylate [7, 8]. In polymer networks from oligomeric macromonomers with a smaller molar mass this influence is more pronounced than for polymer networks obtained from oligomers having a larger molar mass.

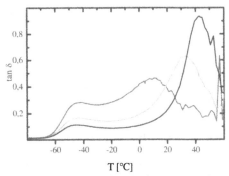

T [°C]

Figure 2. Thermomechanical properties of networks from ABA triblockdimethacrylates determined by DMTA ($M_n$ (NMR) of macrodiols: dark grey 6 kD, light grey 8 kD, black 10 kD) [9].

In the shape-memory polymer network presented above the switching domains were formed by a one segment type. In ABA polymer networks synthesized from poly[(*rac*-lactide)-*block*-polypropyleneglycol-*block*-(*rac*-lactide)]dimethacrylate precursors (PRxtytN) a $T_g$ of the phase provided by the poly(propylene oxide) of -50 °C and a $T_g$ of 50 °C of the phase resulting from the poly(*rac*-lactide) was expected. The thermomechanical properties of three polymer networks from ABA triblockdimethacrylates were determined in dynamic mechanical experiments at varied temperature (DMTA). In these polymer networks a distinct phase separation of the resulting polymer networks could be observed for macrodimethacrylate precursors with $M_n > 10$ kD only. When $M_n$ of the macrodimethacrylate precursors was below 10 kD an additional mixed phase transition between the phase transition resulting from the poly(propylene oxide) and the poly(*rac*-lactide) as well as the phase transition from the poly(propylene oxide) could be detected. The influence of the mixed phase could be observed and is shown in Figure 2.

**Shape-memory properties**

The shape-memory properties of the different polymer network architectures were investigated in cyclic, thermomechanical tests. In Figure 3 the result of such a test, which was performed in strain-controlled mode, is shown for a network from PCL-dimethacrylate by plotting stress as a function of strain. In the first step the sample was loaded and elongated to $\varepsilon_m$ (I). When the strain is kept constant a decrease in stress during lowering of T could be observed, resulting from the entropy elasticity of the amorphous chain segments (II). During cooling to $T_{low}$ crystallization of the switching segments is occurring. During the ongoing cooling process (III) the crystallization results in a dramatic increase of stress to a maximum stress at $T_{low}$ (after thermal contraction; $\sigma_l$).

7

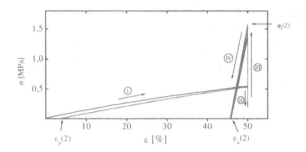

Figure 3. Cyclic thermomechanical, strain-controlled test of a polymer network from PCL-dimethacrylate with $T_h = 70$ °C, $T_l = 0$ °C, and $\varepsilon_m = 50\%$ ($\sigma$ = stress) in air. The data of the graph are obtained from five subsequent cycles, black: first cycle, grey: cycles 2-5 [10].

This increase in stress is related on the one hand to the oriented crystallization of the strained crystalline segments as well as on the other hand on the entropy elastic expansion. Nevertheless, the newly formed crystallites are acting as physical crosslinks which are fixing the tensile-loaded shape of the sample. Upon unloading, the stress-strain curve intersected the $\sigma$ axis at a temporary strain ($\varepsilon_u$), representing the temporary shape of the specimen (IV). A residual strain ($\varepsilon_p$) remained after reheating and recovery of the permanent shape, so that the second cycle (N = 2) was slightly differing from the previous cycle. Generally the permanent shape was recovered by more than 90%, but this value might be increased in subsequent cycles because the preorientation of the sample resulting from processing of the shaped body was deleted. A typical protocol consisted of five cycles.

The mechanical as well as the shape-memory properties of the polymer network having crystalline switching domains could be significantly influenced by the introduction of a second amorphous phase [7, 8]. Here $n$-butyl acrylate was chosen as a second comonomer, which influenced the ratio of crystalline domains below $T_{trans}$ of the switching domains. In these AB polymer networks from PCG-dimethacrylate and $n$-butylacrylate, the glycolide content of the crystalline segments was kept constant as well as the ratio of the $n$-butyl acrylate while the length of the crystallizable segments was varied. Results from strain-controlled thermomechanical experiments cycles of these polymer networks are shown in Figure 4. The curves of the polymer networks N-CG(14)-7 and N-CG(14)-10 were similar to the curve described for the polymer networks purely consisting of crystalline switching segments shown in Figure 3. The curves of the polymer networks N-CG(14)-3 and N-CG(14)-5 differed from the curves described in Figure 3 because the $\varepsilon$-hydroxycaproate-sequences dominated segments were relatively short forming less stable crystallites, stabilizing the temporary shape.

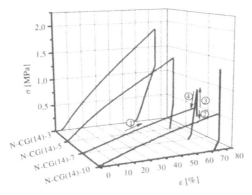

Figure 4. Fifth cycle of a strain-controlled cyclic, thermomechanical tensile test of copolymer networks N-CG(14)-X from macrodimethacrylates with X = 3, 5, 7 or 10 kD of the oligomer in air. $T_h = 70\ °C$, $T_l = 0\ °C$, $\varepsilon_m = 75\%$ [8].

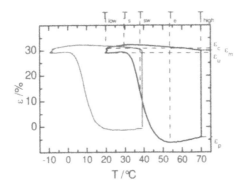

Figure 5. Stress-controlled cyclic, thermomechanical tests of AB copolymer networks from poly[(L-lactide)-ran-glycolide]dimethacrylate and n-ethyl acrylate, (black line: 34 wt% n-ethyl acrylate, grey line: 85 wt% n-ethyl acrylate) [11].

In AB shape-memory polymer networks having an amorphous phase associated to a $T_g$, the second phase, which contributes additional elasticity to the material might also influence the shape-memory properties, when the phase separation between both amorphous phases is not sufficient [11]. In Figure 5 the curves of two cyclic, thermomechanical tests under stress-controlled conditions are presented from polymer networks obtained from copolymerization of poly[(L-lactide)-ran-glycolide]dimethacrylate with n-ethyl acrylate. In such cycles the switching temperature $T_{sw}$ of the thermally-induced shape-memory effect could be determined. $T_{sw}$ was defined as inflection point of the recovery sector of cyclic, thermomechanical tests displayed in Figure 5. The n-ethyl acrylate content of the polymer networks was varied between 34 and 85wt%. From the curves obtained, it could be concluded that $T_{sw}$ was significantly lowered with increasing weight content of n-ethyl acrylate.

A disadvantage of AB polymer networks with an *n*-butyl acrylate content of more than 50 wt-% was a decrease in shape fixity ratio $R_f$. Shape-memory polymer networks, which were elastic at $T < T_{sw}$ and avoided the decreased $R_f$ of AB polymer networks were realized by ABA polymer networks from poly[(*rac*-lactide)-*block*-poly(propylene oxide)-*block*-poly(*rac*-lactide)]dimethacrylate.

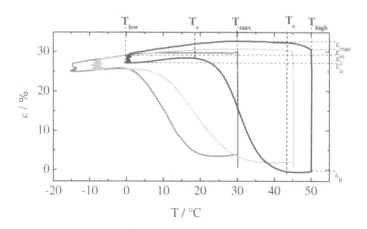

Figure 6. Cyclic thermomechanical tensile experiment at constant stress $\sigma_m$ during cooling and $\sigma = 0$ during reheating for polymer networks from poly[(*rac*-lactide)-*block*-poly(propylene oxide)-*block*-poly(*rac*-lactide)]dimethacrylates. (dark grey: PR4t6tN; light grey PR4t8tN; black: PR4t10tN) [9]

As expected from the DMTA measurements, $T_{sw}$ was closely associated to $T_g$ and was increasing with increasing $T_g$ determined for the ABA polymer networks. The temperature interval of the strain recovery increased between 29 K and 37 K with increasing molecular weight of the chain segments. Additionally, $R_r$ was increasing with increasing molecular weight of the poly(*rac*-lactide) blocks. The chain segments reshaped into the random coil conformation at $T_{sw}$ in the order PR4t6tN < PR4t8tN < PR4t10tN. This behavior could be explained by the increase of the molecular weight of the chain segment per unit volume. In these polymer systems the highest measured entropy gain, which drove the strain recovery, was determined for the block copolymer network PR4t10tN. Consequently, the highest $R_r$ value of 99.5% was measured in case of PR4t10tN. It could be concluded that ABA polymer networks have a suitable polymer network architecture, which provided appropriate elasticity as well as sufficient shape-memory properties. Nevertheless, the phase separation between both phases needed to be sufficient otherwise $T_{sw}$ was closely related to the $T_g$ of the mixed phase. In such a case an independent control of the thermal as well as the shape-memory properties was not given. With increasing $M_n$ of the building blocks sufficient phase separation could be achieved.

## CONCLUSIONS

Design principles of shape-memory polymer networks having crystallizable or non-crystallizable amorphous switching domains were presented. Macrodimethacrylates derived from PCL- or PCG-diols were used for the syntheses of polymer networks with crystallizable switching segments while polymer networks with amorphous switching segments were obtained from PLGA-macrodimethacrylates. Copolymerization of the macrodimethacrylates having either amorphous or crystalline switching segments with various acrylates yielded AB polymer networks. Amorphous ABA polymer networks were obtained by polymerization of macrodimethacrylates from poly[(*rac*-lactide)-*block*-poly(propylene-oxide)-*block*-poly(*rac*-lactide)]-segments.

In polymer networks with amorphous switching domains it was demonstrated that $T_g$ is independent from $M_n$ of the oligomers used as precursors. In contrast, $T_m$ of polymer networks from PCL- or PCG-dimethacrylates with crystallizable switching segments depended on the molecular weight of the oligomers. The elastic properties of shape-memory polymer networks in their permanent and temporary shape were significantly enhanced by the incorporation of a second amorphous phase. In AB polymer networks with $T_{trans} = T_m$ with increasing amount of the amorphous phase a decrease in $R_f$ was observed. In AB polymer networks with $T_{trans} = T_g$, $T_{sw}$ was a function of acrylate content. Polymer networks being elastic at $T < T_{switch}$ and having higher $R_f$ were realized in ABA polymer networks from poly[(*rac*-lactide)-*block*-poly(propylene oxide)-*block*-poly(*rac*-lactide)]dimethacrylate. In these polymer networks a distinct phase separation of the resulting polymer networks could be observed for macrodiol-dimethacrylate precursors with $M_n > 10$ kD only. An additional transition associated to the mixed phase between the phase transition resulting from the poly(propylene oxide) and the poly(*rac*-lactide) as well as the phase transition from the poly(propylene oxide) were observed, when $M_n$ of the macrodiol-dimethacrylate precursors was $< 10$ kD. In conclusion, shape-memory polymer networks with suitable elastic properties above and below $T_{trans}$ could be realized in ABA polymer networks when $M_n$ of the building blocks is sufficiently high and no mixed phase transition is occurring.

In conclusion, the availability of the presented shape-memory polymer network systems and the underlying structure-property relationships enables tailoring of these materials for specific applications.

## REFERENCES

[1] A. Lendlein, S. Kelch, Functionally Graded Materials Viii 492-493 (2005) 219-223.
[2] A. Lendlein, S. Kelch, Angewandte Chemie-International Edition 41 (2002) 2034-2057.
[3] M. Behl, A. Lendlein, Soft Matter 3 (2007) 58-67.
[4] M. Behl, A. Lendlein, Materials Today 10 (2007) 20-28.
[5] C. Liu, H. Qin, P. T. Mather, Journal of Materials Chemistry 17 (2007) 1543-1558.
[6] C. M. Yakacki, R. Shandas, C. Lanning, B. Rech, A. Eckstein, K. Gall, Biomaterials 28 (2007) 2255-2263.
[7] A. Lendlein, A. M. Schmidt, R. Langer, Proceedings of the National Academy of Sciences of the United States of America 98 (2001) 842-847.
[8] S. Kelch, S. Steuer, A. M. Schmidt, A. Lendlein, Biomacromolecules 8 (2007) 1018-1027.
[9] N. Y. Choi, S. Kelch, A. Lendlein, Advanced Engineering Materials 8 (2006) 439-445.
[10] A. Lendlein, A. M. Schmidt, M. Schroeter, R. Langer, Journal of Polymer Science Part a-Polymer Chemistry 43 (2005) 1369-1381.
[11] S. Kelch, N. Y. Choi, Z. Wang, A. Lendlein, Advanced Engineering Materials 10 (2008) 494-502.

Mater. Res. Soc. Symp. Proc. Vol. 1190 © 2009 Materials Research Society    1190-NN01-02

# Effect of chemical crosslinking on the free-strain recovery characteristics of amorphous shape-memory polymers

Alicia M. Ortega[1], Christopher M. Yakacki[2,3], Sean A. Dixon[2,3], Alan R. Greenberg[1], and Ken Gall[2,3,4]

[1]Department of Mechanical Engineering, University of Colorado, Boulder, CO 80309, U.S.A.
[2]MedShape Solutions Inc., Atlanta, GA 30318, U.S.A.
[3]School of Materials Science and Engineering, Georgia Institute of Technology, Atlanta, GA 30332, U.S.A.
[4]Woodruff School of Mechanical Engineering, Georgia Institute of Technology, Atlanta, GA 30332, U.S.A.

## ABSTRACT

The goal of this study is to investigate the fundamental relationship between the extent of crosslinking and shape-memory behavior of amorphous, (meth)acrylate-based polymer networks. The polymer networks were produced by copolymerization of tert-butyl acrylate (tBA) and poly(ethylene glycol) dimethacrylates of differing molecular weights (PEGDMA). Polymer compositions were tailored via the amount (weight percent (wt%)) and molecular weight of the PEGDMA crosslinking agents added to produce four materials with varying levels of crosslinking (0, 2, 10, and 40 wt% crosslinking agent corresponding to 0, 0.6, 3.2, and 16.6 mole%) and nearly equal glass transition temperatures ($T_g$). The effect of crosslinking on deformation limits and free-strain recovery is evaluated. Near complete strain recovery was demonstrated by all materials; however, absolute recovery strain decreased with increasing crosslinking due to a corresponding decrease in strain-to-failure. The results provide insights regarding the link between polymer structure, deformation limits, and strain-recovery capabilities of this class of shape-memory polymers. An improved understanding of this relationship is pivotal for optimizing system response for a wide range of shape-memory applications.

## INTRODUCTION

Shape-memory materials are defined by their ability to recover to an original, permanent shape from a temporary, stored shape with the application of an external stimulus such as an increase in temperature. Shape-memory polymers have been proposed for use in a number of applications such as microfluidic devices [1], stents [2-3], blood clot removal devices [4-5], and orthopedic fixation devices [6], all of which take advantage of the actuation capabilities of these functional materials. Amorphous, chemically crosslinked networks are one class of polymeric materials that have been shown to evidence shape-memory behavior [8]. Listed advantages of this class of shape-memory polymers include good shape recovery, tunable recovery work capacity (by varying crosslinking degree), and no chain slippage [8].

Polymer systems in which the physical, thermomechanical, and shape-memory properties of the polymers could be easily and systematically modified (tuned) have been suggested for use in shape-memory polymer design since they would allow for the control of properties to meet the different requirements associated with a wide range of potential applications [7-8]. Polymer systems based on the copolymerization of (meth)acrylate-based polymers are of considerable interest due to the ease with which their thermomechanical properties can be tailored. Previous studies have shown that thermomechanical properties that should inherently affect the shape-memory behavior of amorphous, chemically crosslinked shape-memory polymers, such as the glass transition temperature ($T_g$), rubbery modulus ($E_R$), and deformation limits, can be controlled by simple changes in comonomer chemistries and concentrations [3, 6-7,9-13]. Additionally, initial studies have shown how aspects of shape-memory behavior of such materials are influenced by such structural modifications in moderately crosslinked networks [3,6,10]. In this investigation, the compositional ratio of dimethacrylate crosslinking agent to acrylate comonomer is varied to systematically alter the level of crosslinking in an amorphous (meth)acrylate-based system. The effect of crosslinking level, ranging from uncrosslinked to highly crosslinked, on strain-to-failure and free-strain recovery behavior is analyzed.

**EXPERIMENT**

Network materials were prepared by the copolymerization of tert-butyl acrylate (tBA) with poly(ethylene glycol) dimethacrylate ($M_n$ = 550 and $M_n$ = 330). A mixture of the two poly(ethylene glycol) dimethacrylate monomers was developed such that after polymerization a $T_g$ equal to that of the homopolymerized tBA was obtained. This PEGDMA monomer mixture will be referred to as the "crosslinking agent" throughout the remainder of the text. The tBA was then copolymerized with varying concentrations of the crosslinking agent (0, 2, 10, and 40 wt% crosslinking agent corresponding to 0, 0.6, 3.2, and 16.6 mole%) to produce four different polymeric materials with similar $T_g$ values (average value of 56 ± 1 °C) and differing crosslinking levels. The photoinitiator 2,2-dimethoxy-2-phenylacetophenone was added to the comonomer solution at a concentration of 1 wt% of the total comonomer weight. Samples (1 mm thick) were photopolymerized with an UV lamp (Blak-Ray, Model B100AP, UVP, Upland, CA). Results of swelling experiments in 2-propanol confirmed the expected increase in crosslink density with increasing crosslinking agent concentration. The homopolymerized tBA polymer dissolves completely when placed in 2-proponol indicating negligible chemical crosslinking. For the remaining materials, the equilibrium swelling ratios in 2-propanol decrease with increasing crosslinking agent concentration, indicating a corresponding increase in crosslink density. Gel contents of the three crosslinked networks are ≥ 93%, indicating near complete incorporation of the initial components in the final three-dimensional network structures. All chemicals were obtained from Aldrich (Milwaukee, WI) and used as-received.

Flat dog-bone tensile samples were used for both strain-to-failure and free-strain recovery shape-memory testing. The tensile samples had a gage cross-sectional area of 3 mm x 1 mm and a gage length of approximately 9 mm. Strain-to-failure and free-strain recovery tests were performed using a mechanical tester (Model 5567, Instron, Norwood, MA) with a 500 N load cell. A thermal chamber (Model 3119-506-A2B3, Instron, Norwood, MA) with liquid nitrogen cooling was used to control the test temperature. For strain-to-failure testing, the materials were tested at $T_g$ (56 ± 1 °C). Tests were run in displacement control at an extension rate of 5 mm/min. Strain was measured externally by a video extensometer (Model 2663-821, Instron, Norwood, MA). Free-strain recovery tests consisted of four steps (Figure 1): (1) the sample was deformed (at $T_g$) to 90% of the strain-to-failure previously determined for the particular composition; (2) the temperature of the sample was decreased from $T_g$ to 20 °C while extension was held constant; (3) the sample was unloaded and constraints were removed; and (4) the temperature was increased at a rate of 2 °C/min (while monitoring strain) until the recovery strain reached a relatively constant value.

14

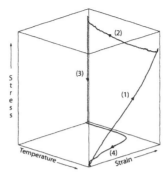

**Figure 1.** Schematic of the free-strain recovery shape-memory cycle: (1) sample is deformed under tensile load (at $T_g$) to 90% of average strain-to-failure; (2) sample is held at constant extension as temperature is reduced to 20 °C; (3) stress is removed from sample; (4) strain of the unconstrained sample is monitored as temperature is increased at a rate of 2 °C/min.

## RESULTS & DISCUSSION

The goal of this work is to determine the fundamental link between crosslinking and the shape-memory behavior of amorphous, (meth)acrylate-based polymers. Specifically, the effect of crosslinking on the deformation limits and free-strain recovery is of interest. In this study, amorphous, chemically crosslinked polymers were produced by the copolymerization of (meth)acrylate-based polymers, and their properties were controlled via changes in comonomer chemistries and concentrations. Comonomer compositions were developed to produce four materials with equivalent $T_g$ values and varying levels of crosslinking, ranging from uncrosslinked (0 wt% crosslinking agent) to highly crosslinked (40 wt% crosslinking agent). The development of four amorphous polymer materials with different degrees of crosslinking and equivalent $T_g$ values enables a direct analysis of the effect of crosslinking on the shape-memory behavior of these materials without complication from varying $T_g$ values and/or storage temperatures.

Strain-to-failure values of these materials at $T_g$ are shown in Figure 2 as a function of crosslinking agent concentration. The strain-to-failure values of these network materials decrease significantly with increasing crosslinking agent concentration. Previous work with a similar system of network polymers evidenced a similar trend with strain-to-failure decreasing rapidly with increasing crosslinking agent concentration and ultimately reaching a plateau in value at high crosslinking agent concentrations [11]. A wide range of strain-to-failure values is captured by this particular system of polymer networks, with average values ranging from 36% to 316%. A strain-to-failure value for the uncrosslinked material (0 wt% crosslinking agent) is not presented since this material does not reach the point of fracture within the extension limits of the tensile tester. We note that the presence of residual monomer in these materials could result in somewhat higher strain-to-failure values. However, such an effect should be greatest in the uncrosslinked material and any such contributions to the strain-to-failure values of the crosslinked materials are likely small given the high gel fractions previously noted. In addition, the presence of polymer chains not incorporated into the network materials would also be expected to increase strain-to-failure values; however, these effects should also be relatively modest. With the strain-to-failure value serving as an upper limit to the recovery strain that a material could potentially evidence, this system shows promise with respect to the 'tailorability' of free-strain recovery shape-memory behavior.

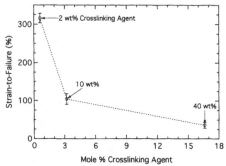

**Figure 2.** Strain-to-failure of the network materials as a function of crosslinking agent concentration. Strain-to-failure at the average glass transition temperature of these materials (56 ± 1 °C) is defined in this study as the tensile engineering strain at which each material fractures.

Free-strain recovery profiles for the four materials are shown in Figure 3. Here the materials have previously been strained to 90% of their average strain-to-failure at $T_g$ and cooled to 20 °C to store the deformation. The absolute strains that are recovered while the unconstrained samples are heated from the storage temperature (20 °C) at a rate of 2 °C/min are shown in Figure 3a. As the materials are heated, they begin to return to their original shapes (i.e. 0% strain). All four compositions demonstrate near complete strain recovery. This is better demonstrated in Figure 3b, where the normalized recovery strains are plotted as a function of temperature. Here normalized recovery strain values of zero and one indicate no recovery and complete recovery, respectively. All four materials approach a normalized recovery strain of one as their temperature is increased, including the uncrosslinked material.

While all materials did demonstrate near complete recovery of their applied strains, there are clear drawbacks and benefits to increasing or decreasing the crosslinking level in these materials with respect to their free-strain recovery shape-memory behavior. As the crosslinking level is increased, the absolute strain that the materials can recover is dramatically decreased (Figure 3a). This is due to the decrease in strain-to-failure of the materials as crosslinking agent concentration is increased (Figure 2). The uncrosslinked and lightly crosslinked materials were able to recover strains on the order of 275%, whereas the most highly crosslinked material is limited to recovery strains on the order of 30%, due to its low strain-to-failure.

The 0, 2, and 10 wt% crosslinking agent materials appear to demonstrate an onset of strain recovery at similar temperatures (slightly above 40 °C), whereas the most highly crosslinked material demonstrates the onset of shape recovery at a slightly lower temperature. Similarly, the most highly crosslinked material reaches full recovery at a lower temperature than the other three materials, followed by the next highly crosslinked material (10 wt% crosslinking agent), and lastly by the most lightly crosslinked and uncrosslinked materials. Similar trends of more highly crosslinked amorphous polymers demonstrating increased speeds of recovery compared to more lightly crosslinked counterparts in free-strain recovery have been noted in previous studies [3,14] and have been attributed to the more highly crosslinked polymers storing more elastic deformation energy and thus increased restorative forces. A lower recovery onset and end temperature could presumably be seen as either a positive or negative design consideration depending upon the final application. Thus, differences in the temperature at which shape-recovery begins and ends should be taken into account when designing a material for specific shape-memory applications.

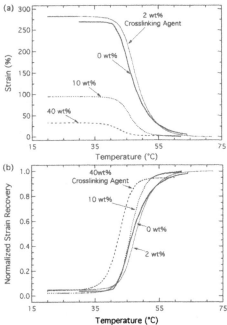

**Figure 3.** Representative free-strain recovery curves of the four polymeric materials used in this study: (a) Absolute tensile strain recovery as a function of temperature; (b) Normalized tensile strain recovery as a function of temperature. Normalized strain recovery is equal to $1 - (\varepsilon / \varepsilon_m)$, where $\varepsilon_m$ is the maximum strain that was reached during material deformation.

## CONCLUSIONS

Crosslinking agent concentration in this (meth)acrylate-based amorphous polymer system was determined to greatly affect free-strain recovery behavior. Polymers of varying levels of crosslinking demonstrated near complete strain recovery of strains on the order of 90% of their strain-to-failure values. The temperature at which the shape recovery was complete is a function of crosslinking agent concentration, and decreases slightly with increasing levels of crosslinking. The temperature at which shape recovery begins also appears related to crosslinking levels, with the most highly crosslinked material showing a shape-recovery onset temperature lower than that of the three more lightly crosslinked materials. The level of absolute strain that these materials can recover decreases with increasing levels of crosslinking due to the corresponding decrease in strain-to-failure values.

Beyond free-strain recovery, the degree of crosslinking should have an effect on the constrained-strain stress-generation behavior of such materials and could also affect storage behavior of the materials over an extended period of time. Further study of the effect of crosslinking on these aspects of shape-memory behavior is currently underway. Overall, this work should help highlight fundamental relationships between crosslinking, deformation limits, strain recovery, and stress generation in amorphous, chemically crosslinked shape-memory polymers. An improved understanding of these relationships is critical for optimizing system response for a wide range of applications including those focused on biomedical device design.

## ACKNOWLEDGMENTS

This project was financially supported in part by Grant Number F31AR053466 from the National Institute of Arthritis and Musculoskeletal and Skin Diseases and the National Science Foundation Alliances for Graduate Education and the Professoriate grant (NSF HRD-0639653) at the University of Colorado. The authors thank Roxanne Likos for assistance with tensile tests.

## REFERENCES

1. K. Gall, P. Kreiner, D. Turner, M. Hulse, Journal of Microelectromechanical Systems **13** (3), 472-483 (2004).
2. K. Gall, C. M. Yakacki, Y. P. Liu, R. Shandas, N. Willett, K. S. Anseth, Journal of Biomedical Materials Research **73A** (3), 339-348 (2005).
3. C. M. Yakacki, R. Shandas, C. Lanning, B. Rech, A. Eckstein, and K. Gall, Biomaterials **28** (14), 2255-2263 (2007).
4. D. J. Maitland, M. F. Metzger, D. Schumann, A. Lee, T. S. Wilson, Lasers Surg Med **30** (1), 1-11 (2002).
5. M. F. Metzger, T. S. Wilson, D. Schumann, D. L. Matthews, and D.J. Maitland, Biomedical Microdevices **4** (2), 89-96 (2002).
6. C. M. Yakacki, R. Shandas, D. Safranski, A. M. Ortega, K. Sassaman, and K. Gall, Advanced Functional Materials **18** (16), 2428-2435 (2008).
7. C. Liu and P. T. Mather, Journal of Applied Medical Polymers **6** (2), 47-52 (2002).
8. C. Liu, H. Qin, and P. T. Mather, Journal of Materials Chemistry **17** (16), 1543-1558 (2007).
9. N. Choi and A. Lendlein, Soft Matter **3** (7), 901-909 (2007).
10. S. Kelch, N. Choi, Z. Wang, and A. Lendlein, Advanced Engineering Materials **10** (5), 494-502 (2008).
11. A. M. Ortega, S. E. Kasprzak, C. M. Yakacki, J. Diani, A. R. Greenberg, and K. Gall, Journal of Applied Polymer Science **110** (3), 1559-1572 (2008).
12. D. Safranski and K. Gall, Polymer **49** (20), 4446-4455 (2008).
13. C. M. Yakacki, S. Willis, C. Luders, and K. Gall, Advanced Engineering Materials **10** (1-2), 112-119 (2008).
14. D. Zhang, X. Lan, Y. Liu, J. Leng, in *Behavior and Mechanics of Multifunctional and Composite Materials*, edited by M. J. Dapino (Proceedings of SPIE **6526**, Bellingham, WA, 2007) pp. 65262W 1-6.

Mater. Res. Soc. Symp. Proc. Vol. 1190 © 2009 Materials Research Society 1190-NN01-04

# A Thermoviscoelastic Approach for Modeling Shape Memory Polymers

Thao D. Nguyen
Department of Mechanical Engineering, The Johns Hopkins University, 3400 N. Charles St., Baltimore, MD 21218

## ABSTRACT

This paper presents a thermoviscoelastic model for shape memory polymers (SMPs). The model has been developed based on the hypothesis that structural and stress relaxation are the primary shape memory mechanisms of crosslinked glassy SMP, and that consideration of these mechanisms is essential for predicting the time-dependence of the shape memory response. Comparisons with experiments show that the model can reproduce the rate-dependent strain-temperature and stress-strain response of a crosslinked glassy SMP. The model also captures many important features of the temperature and time dependence of the free strain recovery and constrained stress recovery response.

## INTRODUCTION

Thermally activated shape memory polymers are active materials that respond to a specific temperature range by changing shape [1]. The permanent shape of a crosslinked, glassy SMP device is manufactured using conventional processes for thermoset polymers, while the temporary shape can be programmed using a thermomechanical cycle as described in [1]. The majority of constitutive models for SMPs treat the shape memory mechanism as a phase transition mechanism (e.g., [2-4]). These models depict the SMP as a mixture of glassy and rubbery phases. The volume fractions of the phases evolve with temperature to produce shape storage and recovery. In this work, we hypothesize that structural and stress relaxation are the primary shape memory mechanisms of crosslinked glassy SMPs. Structural relaxation refers to the time-dependent process in which the macromolecular structure and structure-dependent properties (e.g., the viscosity) evolve to equilibrium in response to a temperature and/or pressure change. Similarly, stress relaxation describes the time-dependent process in which the stress response evolves to equilibrium in response to a change in the deformation state. To examine the relative importance of the structural and stress relaxation mechanisms, we have developed a constitutive model that exhibits structural relaxation in the glass transition region, viscoelasticity in the rubbery and transition regions, and viscoplasticity in the glassy region. A detailed development of the model and discussion of the results below can be found in [5].

## MODEL FORMULATION

Let $\Omega_0$ denote the equilibrium configuration of an undeformed continuum body at time $t_0$ and temperature $T_0$, and $\mathbf{F}$ denote the deformation gradient that maps $\Omega_0$ to the heated/cooled deformed configuration $\Omega$. It is assumed that the deformation gradient can be split multiplicatively into thermal and mechanical components, $\mathbf{F} = \mathbf{F}_M \mathbf{F}_T$, where $\mathbf{F}_T = \Theta_T^{1/3} \mathbf{1}$ and $\Theta_T$ is the thermal dilatation. The mechanical deformation gradient is split further into elastic and viscous parts, $\mathbf{F}_M = \mathbf{F}^e_M \mathbf{F}^v_M$. We also assume that the mechanical deformation gradient and its components can be split multiplicatively into deviatoric and volumetric parts as, $\mathbf{F}_M = \Theta_M^{1/3} \overline{\mathbf{F}}_M$. Then, the deviatoric right and left deformation tensors can be defined as $\overline{\mathbf{C}}_M = \Theta_M^{-2/3} \mathbf{F}_M^T \mathbf{F}_M$,

$$\overline{\mathbf{C}}^e_M = \Theta_M^{e-2/3} \mathbf{F}_M^{eT} \mathbf{F}^e_M, \quad \overline{\mathbf{b}}_M = \Theta_M^{-2/3} \mathbf{F}_M \mathbf{F}_M^T \text{ and } \overline{\mathbf{b}}^e_M = \Theta_M^{e-2/3} \mathbf{F}^e_M \mathbf{F}_M^{eT}.$$

To model the structural relaxation response to a temperature change, we assume that the thermal dilatation is small and that it can be split additively as follows:

$$\Theta_T\left(T,T_f\right) = 1 + \alpha_g\left(T - T_0\right) + \Delta\alpha\left(T_f - T_0\right) \tag{1}$$

The parameter $\alpha_r$ is the coefficient of thermal expansion (CTE) of the rubbery material, $\alpha_g$ is the CTE of the glassy material, and $\Delta\alpha = \alpha_r - \alpha_g$. This formulation splits the thermal response into an instantaneous part that scales with the temperature change and a part signifying the departure from equilibrium $\overline{\delta}^{neq} = \Delta\alpha\left(T_f - T_0\right)$. The fictive temperature $T_f$ is a concept developed by Tool [6] to model structural relaxation of annealing glasses. We apply the Tool model for $T_f$ and the nonlinear Adam-Gibbs model [7-9] to develop an evolution law for $\overline{\delta}^{neq}$:

$$\dot{\overline{\delta}}^{neq} = -\frac{1}{\tau_R}\left(\overline{\delta}^{neq} - \Delta\alpha(T - T_0)\right), \quad \tau_R = \tau_R^{ref}\exp\left[-\frac{C_1}{\log e}\left(\frac{C_2(T - T_f) + T(T_f - T_g^{ref})}{T(C_2 + T_f - T_g^{ref})}\right)\right], \tag{2}$$

where $\overline{\delta}^{neq}(0) = 0$. The parameter $\tau_R$ is the structural relaxation time that depends on $T$ and $T_f$, $\tau_R^{ref}$ is the viscosity at $T_g^{ref}$, which is the glass transition temperature measured at a reference cooling rate, and $C_1$ and $C_2$ are the Williams-Landel-Ferry (WLF) constants for the time-temperature shift factor. Using eq. (2), it can be shown that $T_f$ evolves to its equilibrium value, $T$, for $T >> T_g$. Then, the expression for $\tau_R$ reduces to the WLF equation for the temperature dependence of the relaxation time of a rubbery polymer. For $T << T_g$, the fictive temperature becomes a constant, $T_g$, and $\tau_R$ reduces to the Arrhenius equation for the temperature dependence of a glassy polymer.

To model viscoelastic stress relaxation, it is assumed that the stress response can be additively split into equilibrium and nonequilibrium deviatoric components and a pressure component, $\sigma = s^{eq} + s^{neq} + p\mathbf{1}$. The former represents the low-temperature, rubbery response while the latter dominates the low-temperature, glassy response. Specifically, we use the Arruda-Boyce model [10] for the equilibrium deviatoric stress response, and the compressible Neo-Hookean model for the nonequilibrium stress response:

$$s^{eq} = \frac{1}{J}\mu^{eq}\left(\overline{\mathbf{b}}_M - \frac{1}{3}\overline{I}_{M1}\mathbf{1}\right), \quad \mu^{eq} = \mu_N\left(\frac{\lambda_L}{\overline{\lambda}_{eff}}\right)L^{-1}\left(\frac{\overline{\lambda}_{eff}}{\lambda_L}\right)$$

$$s^{neq} = \frac{1}{J}\mu^{neq}\left(\overline{\mathbf{b}}_M^e - \frac{1}{3}\overline{I}_{M1}^e\mathbf{1}\right), \quad p = \frac{1}{J}\kappa(\Theta_M - 1). \tag{3}$$

The parameters $\overline{I}_{M1} = \overline{\mathbf{b}}_M : \mathbf{1}$ and $\overline{I}_{M1}^e = \overline{\mathbf{b}}_M^e : \mathbf{1}$ are the first invariants of the deformation tensors, $\overline{\lambda}_{eff} = \frac{1}{3}\sqrt{\overline{I}_{M1}}$ is the effective chain stretch, and the function $L^{-1}$ is the inverse Langevin function. The parameters $\mu_N$ is the shear stiffness of the polymer network, $\lambda_L$ is the limiting chain stretch, $\mu^{neq}$ is the nonequilibrium shear modulus, and $\kappa$ is the bulk modulus.

The following evolution equations are developed for the internal deformation $\mathbf{b}_M^e$ based on the Eyring model for viscous flow [11] and for an internal variable $s_y$, which signifies the temperature dependent yield strength of the glassy material:

$$-\frac{1}{2}\left(L_v\mathbf{b}_M^e\right)\mathbf{b}_M^{e^{-1}} = \dot{\gamma}^v\frac{s^{neq}}{\|s^{neq}\|}, \quad \dot{\gamma}^v = \frac{Ts_y}{\sqrt{2}Q_s\eta_s^{ref}}\exp\left[\frac{C_1}{\log e}\left(\frac{C_2(T - T_f) + T(T_f - T_g^{ref})}{T(C_2 + T_f - T_g^{ref})}\right)\right]\sinh\left(\frac{Q_s\|s^{neq}\|}{T\sqrt{2}s_y}\right),$$

$$\dot{s}_y = h\left(1 - \frac{s_y}{s_{y_{ss}}}\right)\dot{\gamma}^v, \quad s_y(0) = s_{y0}, \tag{4}$$

where, $\mathbf{b}_M^e(0) = \mathbf{1}$. We have assumed that the viscous flow rate $\dot{\gamma}^v$ exhibits the same temperature dependence as the structural relaxation time. Following the Eyring model, the viscous flow rate also depends on the magnitude of the flow stress $\|\mathbf{s}^{neq}\| = \sqrt{\frac{1}{2}\mathbf{s}^{neq} : \mathbf{s}^{neq}}$. The parameter $Q_S$ is the activation energy, $h$ is the softening modulus, and $s_{yss}$ is the steady-state yield strength.

## RESULTS AND DISCUSSION

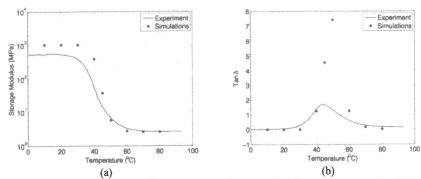

(a)                    (b)

Figure 1. Temperature dependence of the (a) storage modulus and (b) tanδ comparing the results of DMA experiments (at 1Hz and 1°C cooling rate) and finite element simulations[1].

The model was implemented in a finite element program and applied to simulate the thermomechanical experiments and shape recovery experiments for tBA-co-pegdma, a crosslinked, glassy SMP synthesized by polymerizing tert-butyl acrylate monomers with polyethylene glycol dimethacrylate crosslinkers (see Qi et al. [3]). The parameters of the model were determined from a suite of dynamic mechanical analysis (DMA) and isothermal uniaxial compression tests. A detailed description of the finite element simulation and parameter determination procedure can be found in [2]. Figure 1 shows the temperature dependence of the storage modulus and tanδ obtained from the model and from experimental data for the tBA-co-pegdma SMP. The model was able to reproduce temperature dependence of the DMA measurements. There were discrepancies between the model and DMA data regarding the glassy modulus and the peak of the tanδ. The nonequilibrium shear modulus, $\mu^{neq}$, was fitted to the glassy modulus measured by uniaxial compression tests at $T=20°C$. The modulus measured by the uniaxial compression test was twice that measured by the DMA experiments. The peak tanδ calculated from the simulations was four times greater than the data. The model assumed a single stress relaxation process with a single characteristic stress relaxation time $\tau_S = \eta_S / \mu^{neq}$. In contrast, most polymers exhibit a broad spectrum of relaxation times. This simplifying assumption most likely led to the significantly higher peak in the tanδ.

---

[1] Reprinted from *Journal of Mechanics and Physics of Solids*, T.D. Nguyen, H.J. Qi, F. Castro, K.N. Long, "A thermoviscoelastic model for amorphous shape memory polymers: Incorporating structural and stress relaxation", pp. 2792-2814, 2008, with permission from Elsevier.

21

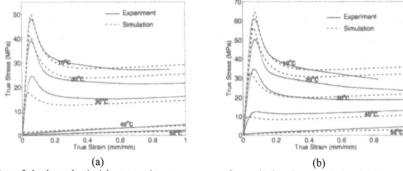

(a)                                                                              (b)

Figure 2. Isothermal uniaxial compression stress response for applied engineering strain rates (a) $\dot{\varepsilon} = 0.01/s$ and (b) $\dot{\varepsilon} = 0.1/s$ comparing simulations and experiments[2].

Figure 2 compares the plots of the isothermal, uniaxial compression stress response at different temperatures and for two strain rates $\dot{\varepsilon} = 0.01/s$ and $\dot{\varepsilon} = 0.1/s$. The parameters $Q_S$, $h$, $s_{y0}$, and $s_{yss}$, were determined by fitting the model to the uniaxial compression data obtained at $T=0°C$ and $T=10°C$ and $\dot{\varepsilon} = 0.01/s$. The results from the simulations showed good agreement with the data for all of the temperatures and strain rates examined, not just those used to determine the model parameters. Specifically, it was able to reproduce the temperature transition of the inelastic behavior from a glassy viscoplastic response, with a distinct yield point and post-yield softening, to a rubbery viscoelastic behavior.

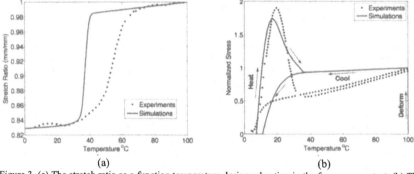

(a)                                                                              (b)

Figure 3. (a) The stretch ratio as a function temperature during reheating in the free recovery test; (b) The normalized engineering stress as a function of temperature in a constrained recovery test[3].

Figure 3a plots the stretch ratio as a function of temperature comparing the experimental measurements and finite element simulation results of the free recovery of a 0.5cm radius by1.0 cm tall plug of the tBA-co-peg material. The plug was deformed at 100°C at $\dot{\varepsilon} = 0.01/s$, cooled at

[2] Ibid.
[3] Ibid.

a rate of $q=1°C/s$ and reheated at the same rate. The model was able to predict the onset of strain recovery at $T_g$, as observed in experiments. However, it predicted a significantly faster recovery time. Full recovery was achieved at a lower temperature in the simulation than in experiments

Figure 3b shows the normalized engineering stress response comparing data and finite element simulation of a constrained recovery experiment on a similar SMP plug. The constrained recovery experiments used the same deformation rate and cooling and heating rates as the unconstrained recovery experiments. The stress was normalized by the stress at the end of the deformation phase of the thermomechanical cycle. The simulation was able to predict many features of the experimental data, including the hysteresis of the stress-temperature response, the normalized peak stress of the reheating curve, and the temperature at which the heating curve joined the cooling curve.

## CONCLUSIONS

A thermoviscoelastic model was developed for crosslinked glassy SMPs that incorporated the effects of structural relaxation and stress relaxation, in the form of viscoelasticity in the glass transition and rubbery regions and viscoplasticity in the glassy region. The model neglected the effects of pressure on structural relaxation, and the effects of heat transfer on the shape memory response. These were negligible in the experiments examined here. The model was able to reproduce the data from DMA and isothermal uniaxial compression tests. More importantly, it was able to predict many important features of the free and constrained recovery response. These results supported the hypothesis that structural and stress relaxation are the primary shape memory mechanisms of crosslinked glassy SMP. However, the model predicted a significantly faster recovery time during the free recovery. This discrepancy was caused most likely by the assumption of a single stress relaxation process. We are currently modifying the model by incorporating a broader relaxation spectrum to test this assertion.

## ACKNOWLEDGMENTS

This work was funded by the National Science Foundation (CMMI-0758390) and the Laboratory Directed Research and Development program at Sandia National Laboratories. Sandia is a multiprogram laboratory operated by Sandia Corporation, a Lockheed Martin Company, for the United States Department of Energy under contract DE-ACO4-94AL85000.

## REFERENCES

1. A. Lendlein, S. Kelch, K. Kratz, J. Schulte, "Shape-memory Polymers", In: *Encyclopedia of Materials: Science and Technology*, (Elsevier 2005), pp. 1–9.
2. H.J. Qi ,T.D. Nguyen, , F. Castro, C. Yakacki, R. Shandas, *J. Mech. Phys. Solids*, **56**, 1730-1751 (2008).
3. Y.Liu, K. Gall, M.L. Dunn, A.R. Greenberg, J. Diani, *Int. J. Plasticity*, **22**, 279–313, (2006).
4. Y-.C. Chen and D.C. Lagoudas, *J. Mech. Phys. Solids*, **56**, 1752-1765 (2008).
5. T.D. Nguyen, H.J. Qi, F. Castro, K.N. Long, *J. Mech. Phys. Solids*, **56**, pp. 2792-2814 (2008).
6. A.Q. Tool, *J. Am. Ceram. Soc.*, **29**, 240–253 (1946).
7. G.W. Scherer, *J. Am. Ceram. Soc.*, **67**, 504 (1984).
8. I. Hodge, *J. of Res. of the NIST*, **102**, 195–205 (1997).
9. G. Adam, and J.H. Gibbs, *J. Chem. Phys.*, **43**, 139-146 (1965).
10. E.M. Arruda, and M.C. Boyce, *J. Mech. Phys. Solids*, **41**, 389–412 (1993).
11. H. Eyring, *J. Chem. Phys.*, **4**, 283-291 (1968)

# Mechanical Properties of Polymer Blends Having Shape-Memory Capability

Marc Behl[1], Ute Ridder[2], Wolfgang Wagermaier[1], Steffen Kelch[3] and Andreas Lendlein[1]
[1]Center for Biomaterial Development, Institute of Polymer Research, GKSS Research Center Geesthacht GmbH, Kantstr. 55, 14513 Teltow, Germany
[2]Freudenberg Forschungsdienste KG, 69465 Weinheim, Germany
[3]Sika Technology AG, Tüffenwies 16, CH-8048 Zürich, Switzerland

## ABSTRACT

The general design principle of shape-memory polymers (SMP) requires two key components: covalent or physical crosslinks (hard domains) determining the permanent shape and switching domains fixing the temporary shape as well as influencing the switching temperature $T_{sw}$. In conventional thermoplastic SMP hard and switching domains determining segments are combined in one macromolecule, e.g. block copolymers such as polyurethanes. Recently, binary polymer blends having shape-memory properties, from two different multiblock copolymers have been presented, whereby the first one is providing the segments forming hard domains and the second one the segments forming the switching domains. Besides the shape-memory properties, the mechanical properties of such materials are application relevant. Here we investigate how the blend composition influences mechanical properties of this new class of shape-memory materials.

## INTRODUCTION

Thermoplastic shape-memory polymers are consisting of hard- and switching segments, which are linked by covalent bonds and form phase-segregated domains [1, 2]. By variation of different molecular parameters, for instance the weight ratio of hard to switching segment, these multiblock copolymers can be designed as polymer systems, in which mechanical and thermal properties can be adjusted over a wide range [3, 4]. This concept was extended to multifunctional polymer systems that are biodegradable and have a shape-memory capability [5, 6]. However, any new combination of material properties in this system requires synthesis of a new material. Recently, binary polymer blends from a first component providing the hard domains and a second component contributing the switching domains were presented. In this way a polymeric shape-memory material was obtained, in which hard and switching segments were not covalently linked [7]. A poly(alkylene adipate) is incorporated as mediator segment in both multiblock copolymers to promote their miscibility as the hard segment poly(p-dioxanone) (PPDO) and the switching segment poly(ε-caprolactone) (PCL) are non-miscible. The binary polymer blends were prepared by co-precipitation from solution as well as by co-extrusion. Shape-memory properties of all investigated polymer blends were excellent. The melting point associated to the PCL switching domains $T_{m,PCL}$ was almost independent from the weight ratio of the two blend components. In this way complex synthesis of new materials can be avoided. Its biodegradability, the variability of mechanical properties and a $T_{sw}$ around body temperature are making this binary blend system an economically efficient, suitable candidate for diverse biomedical applications.

As the understanding of the mechanical properties of this new class of polymeric shape-memory materials is of fundamental interest, the mechanical behavior of the polymer blends at varied temperatures were investigated and will be discussed. The discussion will be focused on polymer blends from co-precipitation. The mechanical properties of the binary polymer blends were investigated by means of tensile tests at 20 °C and 50 °C, which is below and above $T_{trans}$. Below $T_{trans}$ the mechanical properties determine the elastic behaviour e.g. for programming by cold-drawing as well as for the fixed temporary and the permanent shape after the thermally-induced shape-memory effect. Above $T_{trans}$ of the switching segment the elastic properties are important for the programming of the shape-memory effect.

## EXPERIMENTS

The multiblock copolymers providing the hard segment for the polymer blends, named as PDA, were synthesized from poly($p$-dioxanone)diol with a number average molecular weight $M_n$ of 2,800 g•mol$^{-1}$, the poly(alkylene adipate) diol (PADOH) with $M_n$ of 1,000 g•mol$^{-1}$ and a mixture of 2,2,4 and 2,4,4 trimethylhexamethylene diisocynanate (TMDI). Copolyesterurethanes denoted as PCA were prepared from PCL-diol with $M_n$ of 2.000 g•mol$^{-1}$, PADOH, and TMDI, and are providing the switching segment. Two polymers differing in the weight ratio of the two segments were synthesized from each type of multiblock copolymer PDA and PCA [the number in brackets gives the weight content of PPDO blocks in PDA and of PCL blocks in PCA]. For blend formation by precipitation PDA and PCA were dissolved in 1,2-dichlorethane (10 wt% solutions) and then precipitated in a tenfold excess of hexane fraction. The resulting polymer blend was collected by filtration and dried in vacuum until constant weight was reached. Experimental details about sample preparation as well as differential scanning calorimetry experiments (DSC) and wide angle X-ray diffraction (WAXD) were presented in [7]. Mechanical properties at 20 °C and 50 °C were assessed by tensile tests on a ZWICK1425 equipped with a load cell capable for forces up to 50 N combined with a thermo chamber (Climatic Systems LTD, model 091250, controller unit Eurotherm 902-904 unit). The elongation rate was 10 mm•min$^{-1}$. The samples were equilibrated for 20 min at the operating temperature before each experiment. Bone-shaped samples with dimensions of 10 mm × 3 mm (parallel area) were used with a free length between 4 and 6 mm.

## DISCUSSION

Table 1: Thermal and mechanical properties of multibock copolymers PDA and PCA.

| Multiblock copolymer ID[a] | $T_m$[b] (°C) | $\Delta H_m$[b] (J•g$^{-1}$) | $\Delta H_{m,i}$[c] (J•g$^{-1}$) | $E$[d] (MPa) | $\sigma_b$[d] (MPa) | $\varepsilon_b$[d] (%) | $E$[e] (MPa) | $\sigma_b$[e] (MPa) | $\varepsilon_b$[e] (%) |
|---|---|---|---|---|---|---|---|---|---|
| PDA(42) | 90 | 6.3 | 15.0 | 16.5 ± 0.5 | 14.8 ± 1.5 | 1070 ± 60 | 12.0 ± 1.4 | 4.5 ± 0.7 | 690 ± 135 |
| PDA(50) | 91 | 20.4 | 58.2 | 21.9 ± 4.3 | 15.0 ± 1.1 | 930 ± 190 | 21.6 ± 2.8 | 10.6 ± 1.1 | 1220 ± 120 |
| PCA(47) | 37 | 1.7 | 3.6 | 7.7 ± 0.6 | 9.5 ± 1.6 | 1455 ± 145 | n.d. | n.d. | n.d. |
| PCA(68) | 37 | 27.5 | 40.4 | 22.6 ± 7.8 | 15.6 ± 0.5 | 1220 ± 425 | n.d. | n.d. | n.d. |

a) PDA is prepared from PPDO, PADOH and TMDI; two-digit number in brackets gives weight of PPDO segments. PCA is prepared from PCL, PADOH and TMDI; number in brackets is weight content of PCL segment.
b) $T_m$: melting temperature, and $\Delta H_m$: melting enthalpy obtained from DSC second heating run.
c) partial melting enthalpy $\Delta H_{m,i}$; for PDA $\Delta H_{m,PPDO}$; for PCA $\Delta H_{m,PCL}$
d) E: Young's modulus, $\sigma_b$: stress at break, and $\varepsilon_b$: strain at break determined by tensile tests at 20 °C.
e) determined by tensile tests at 50 °C. n.d. not determined, because polymer in melt.

The thermal and mechanical properties of the block copolymers used for blend preparation above and below T_trans are given in table 1. While in the block copolymers providing the hard segments (PDA) nearly no influence of the different temperature conditions can be seen, in the block copolymers providing the switching segment the elastic properties at 50 °C can not be determined due to melting. From the block copolymers polymer blends were prepared according to the experimental section.

The crystallinity of the block copolymers as well as the polymer blends was investigated by means of WAXD (Fig. 1). PDA(50)/PCA(47)[25/24/37] showed in WAXD investigations two characteristic crystalline peaks at 21.33° and 21.77° attributed to the (110) reflection of the PCL crystalline domain and to the (210) reflection of the PPDO crystalline domain, respectively. When the measurement temperature was 50 °C, the 21.33° peak was not observed, while the peaks attributed to PPDO crystalline domain were similar to PDA(50) without any change. After cooling to room temperature again, the peak at 21.33° appeared in the diffraction pattern and the intensity increased with cooling time, indicating the formation of independent crystalline structures by PCL and PPDO.

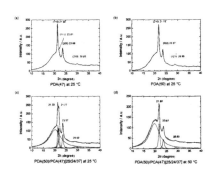

Figure 1. WAXS diffractogram of multiblock copolymers (a) PCA(47) and (b) PDA(50) as well as polymer blend (c,d) PDA(50)/PCA(47)[25/24/37]

The mechanical properties are a function of the degree of crystallinity (DOC) as besides the crystalline phases the amorphous phases of the semicrystalline segments have to be considered [8]. The crystallites are acting as physical crosslinks and increase the stiffness of the binary polymer blends. For the correct determination of DOC of the polymer blends by WAXS the diffraction capability of PCL and PPDO crystal populations has to be the same. As this is not the case, a crystallinity rate (CR) can be estimated by means of DSC measurements using the partial melting enthalpies. When both semicrystalline segments contribute to CR as this is the case at 20 °C, $CR_{20}$ can be estimated according to equation 1, while at 50 °C only the crystalline segments of the PPDO phase have to considered (equation 2).

$$CR_{20}\% = \left( \frac{\Delta H_{m,PPDO}}{\Delta H^0_{PPDO}} + \frac{\Delta H_{m,PCL}}{\Delta H^0_{PCL}} \right) \cdot 100 \quad (1)$$

$$CR_{50}\% = \frac{\Delta H_{m,PPDO}}{\Delta H^0_{PPDO}} \cdot 100 \quad (2)$$

$\Delta H_{m,i}$ is the experimentally determined partial melting enthalpy of the PPDO- and PCL-phase, $\Delta H^0_i$ the melting enthalpy of 100 % crystalline PPDO (140 J·g⁻¹) or PCL (139,5 J·g⁻¹).

**Mechanical properties of polymer blends at 20 °C**

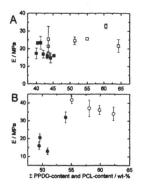

Figure 2.     Young's modulus E of binary polymer blends prepared by co-precipitation as a function of weight content of the semicrystalline segments at 20 °C depending on the overall weight content of PPDO and PCL determined by tensile tests. A polymer blends ■ PDA(42)/PCA(47)and □ PDA(42)/PCA(68), B polymer blends ● PDA(50)/PCA(47)and ○ PDA(50)/PCA(68).

At 20 °C the binary polymer blends have to be considered semicrystalline materials as the $T_m$s associated to the domains providing the switching segments and the domains providing the hard segment are higher than 20 °C. In Figure 2 of the Young's modulus as a function of the weight content of the crystalline segments is presented. In general, the Young's modulus increases with increasing content of semicrystalline segments from 12 MPa to 42 MPa. The Young's modulus of the binary blends PDA(42)/PCA(47), PDA(42)/PCA(68) (Fig. 1 A) and PDA(50)/PCA(47) (Fig. 1  B) are varying around constant values of 18 MPa, 24 MPa and 37 MPa, while the Young's modulus of the blend PDA(50)/PCA(68) (Fig. 1 B) rises with increasing content of the crystalline domains from 13 MPa to 32 MPa.

In Fig. 3 I the Young's modulus as a function of $CR_{20}$ is illustrated. Fig. 3 I A represents the binary polymer blends containing PDA42, while the Young's modulus of the blends containing PDA50 is shown in Fig. 3 I B. A dependency of Young's modulus as a function of $CR_{20}$ was determined: the Young's modulus of the binary polymer blends varied between 12 MPa and 42 MPa having $CR_{20}$ between 1% and 16% and increased with increasing CR, with the exception of PDA(50)/PCA(68), having an Young's modulus between 34 MPa and 42 MPa. The Young's modulus of the polymer blends, which contain PDA(42) as hard segment providing domains, varied between 15 MPa and 33 MPa. Elongation at break ($\varepsilon_b$), which is presented in Fig. 3 as a function of $CR_{20}$, decreased with increasing CR, as the mobility of the chain segments got confined by the crystallites acting as physical crosslinks. $\varepsilon_b$ varied between 440% and 1510%: In both series a decrease of $\varepsilon_b$ with increasing CR could be determined. In binary polymer blends containing PDA(42) (Fig 3 II A) $\varepsilon_b$ values between 1000% and 550% were determined at CR between 8% and 17%. $\varepsilon_b$ of polymer blends containing PDA(50) reached values between 1050% and 600% at CR between 5% and 16%. In case of PDA(50) higher values of $\varepsilon_b$ compared to the PDA-Polymer were determined and could be explained by the elasticity of the second component, the PCA-polymer segment. Stress at break ($\sigma_b$) at 20 °C as a function of $CR_{20}$ is presented in Fig. 3 III. $\sigma_b$ showed the same characteristics as for $\varepsilon_b$ for blends containing PDA(42) and PDA(50). While for binary polymer blends containing PDA(42) $\sigma_b$ varied between 5 MPa and 12 MPa, values for $\sigma_b$ were determined between 9 MPa and 16 MPa for polymer blends containing PDA(50).

28

I                       II                      III

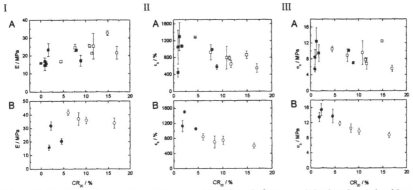

Figure 3. Mechanical properties of binary polymer blends from precipitation determined by tensile test at 20 °C as a function of $CR_{20}$. (I) Young's modulus E, (II) strain at break $\varepsilon_b$, (III) stress at break $\sigma_b$; (A) polymer blends ■ PDA(42)/PCA(47) and □ PDA(42)/PCA(68), (B) polymer blends ● PDA(50)/PCA(47) and ○ PDA(50)/PCA(68).

## Mechanical Properties of polymer blends at 50 °C

At 50 °C the PCL-phase is amorphous, so that only the crystallites of the PPDO domains are acting as physical crosslinks. As a consequence, the mechanical properties are determined by the PPDO-phase only. Resulting from the loss of crystallinity, it is expected that Young's modulus as well as $\sigma_b$ will decrease while $\varepsilon_b$ should increase. To proof this, the binary polymer blends were investigated by means of tensile tests at 50 °C and results are presented in Fig 4. In Fig 4 I Young's modulus as a function of $CR_{50}$ is displayed. As expected, Young's modulus increased with increasing CR. In binary polymer blends containing PDA(42) CR was between 0.5% and 10% and Young's moduli between 0.3 MPa and 17 MPa were determined. In binary polymer blends based on PDA(50) Young's moduli reached values between 1.7 MPA and 18 MPa at increasing CR between 1.8% and 4.5%. $\sigma_b$ is presented in Fig. 4 III and was decreasing with increasing $CR_{50}$. Values between 0.3 MPa and 5 MPa were obtained. A similar behaviour was observed for $\varepsilon_b$ which decreased with increasing CR (Fig. 4 II). In general, when CR was below 5% $\varepsilon_b$ of 200% can be obtained. Polymer blends containing PDA(42) (Fig. 4. II A) reached $\varepsilon_b$ around 60% at $CR_{50}$ of 8%. The effect of an increase of $\varepsilon_b$ was not determined and was attributed to the fact that the loss of mechanical stability overrides the effect of lowered crystallinity.

## CONCLUSIONS

For the shape-memory effect, two major requirements have to be considered: sufficient elastic properties in the range of actuation and the existence of distinct switching and hard domains. Both can be fulfilled by the polymer blend depending on its composition. Mechanical properties of the polymer blends were adjusted by CR and choice of the multiblock copolymers. Young's moduli were varied between 10 and 42 MPa having $\sigma_b$ between 5 MPa und

16 MPa. Depending on CR $\varepsilon_b$ varied between 24% and 1050%. When mechanical measurements were performed above $T_{m,PCL}$, the mechanical properties were only determined by CR of the PPDO phase. Elasticity of the materials decreased significantly when CR increased above 5%. Compared with SMP based on multiblock copolymers of PCL, PPDO and TMDI in which the hard and the switching segment were located in the same multi block polymer [5], the SMP blends are only capable of lower Young's moduli and stresses at break. Nevertheless, the ease of varying mechanical properties by variation of blends' composition, make these shape-memory polymer blends an interesting alternative. In a prospective view, reinforcement of the polymer blends with inorganic particles should allow higher Young's moduli.

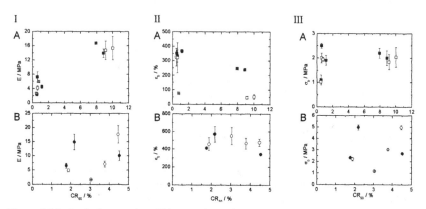

Figure 4. Mechanical properties of binary polymer blends from precipitation determined by tensile test at 50 °C as a function of $CR_{50}$. (I) Young's modulus E, (II) strain at break $\varepsilon_b$, (III) stress at break $\sigma_b$; (A) polymer blends ■ PDA(42)/PCA(47)and □ PDA(42)/PCA(68), (B) polymer blends ● PDA(50)/PCA(47)and ○ PDA(50)/PCA(68).

**REFERENCES**

[1]     A. Lendlein, S. Kelch, *Angewandte Chemie-International Edition* **2002**, *41*, 2034.

[2]     M. Behl, A. Lendlein, *Soft Matter* **2007**, *3*, 58.

[3]     A. Alteheld, Y. K. Feng, S. Kelch, A. Lendlein, *Angewandte Chemie-International Edition* **2005**, *44*, 1188.

[4]     J. Hu, *Shape memory polymers and textiles*, Woodhead Publishing Limited, Cambridge; England **2007**.

[5]     A. Lendlein, R. Langer, *Science* **2002**, *296*, 1673.

[6]     Y. K. Feng, M. Behl, S. Kelch, A. Lendlein, *Macromolecular Bioscience* **2009**, *9*, 45.

[7]     M. Behl, U. Ridder, Y. Feng, S. Kelch, A. Lendlein, *Soft Matter* **2009**, *5*, 676.

[8]     J. M. G. Cowie, S. Harris, J. L. G. Ribelles, J. M. Meseguer, F. Romero, C. Torregrosa, *Macromolecules* **1999**, *32*, 4430.

Mater. Res. Soc. Symp. Proc. Vol. 1190 © 2009 Materials Research Society    1190-NN01-08

# Relationship Between Materials Properties and Shape Memory Behavior in Epoxy-Amine Polymers

Ingrid A. Rousseau[*] and Tao Xie
General Motors Corp., Technical R&D Center, Materials and Processes Lab., 30500 Mound Road, M.C.: 480-106-710, Warren MI 48090-9055, U.S.A.
[*]Corresponding author: ingrid.rousseau@gm.com

## ABSTRACT

Although epoxy-based polymers remain infrequently used as shape memory polymers (SMP's), they are a promising base material for highly demanding applications due to their intrinsic physical properties and ease of processing. A series of epoxy SMP's was synthesized with varying mechanical properties and with glass transition temperatures ranging from 31 to 93 °C, tunable via the variations of the molecular structures. The influence of chemical structures and physical properties of these epoxy SMP's on their shape memory (SM) behavior is examined in detail along with the impact of the shape memory cycling conditions. While the results show that lower crosslink densities and/or higher molecular flexibility/mobility leads decreased SM performance, at low crosslink density the effect of molecular flexibility/mobility becomes dominant in influencing the SM response.

## INTRODUCTION

Shape memory polymers (SMP's) have attracted much interest over the past decade as alternative solutions to other smart materials. These polymers can be controlled to exist in either a *permanent* or *temporary* shape by the combination of an imposed deformation and a triggering stimulus. While SMP's low cost, easy processing, and low density, amongst other qualities, made them materials of choice for the development of new technologies [1-6], their lower mechanical and thermal properties, and their reduced durability compared to their metal alloys analogues (i.e., shape memory alloys), require essential improvements [7-11].

Because glassy thermoset SMP's can be expected to perform better than other SMP classes [11], and because the development of epoxy-SMP's has remained infrequent, we chose to study the SM behavior of epoxy thermosets and the impact of their molecular structures (crosslink density, crosslinker functionality, and chain flexibility) on their shape memory effect. Their shape memory performance was quantified under varying experimental conditions (i.e., deformation load, recovery heating rate, number of SM cycles, holding times in the deformed or temporary shape above and below their transformation temperature).

## EXPERIMENT: SHAPE MEMORY CYCLE CHARACTERIZATION

The shape memory (SM) cycles were measured using the Q800 DMA. The values for the setting temperature ($T_s$), the deformation temperature ($T_d$), the transformation temperature ($T_{trans}$), and the storage modulus at $T_s$ ($E'_s$) and at $T_d$ ($E'_d$) were determined from the thermo-mechanical data [12]. The definition and quantification of the SM performance of a material were discussed in greater details in previous reports [11, 12]. The response temperature ($T_r$), the shape fixity ($R_f$), the shape recovery ($R_r^{min}$ and $R_r^{eu}$), the recovery speed ($V_r$) and recovery time ($t_r$), and the overall SM cycle time ($t_{sm}$) were evaluated.

The *instantaneous* shape memory effect was evaluated by performing a simple SM cycle during which the deformation step at $T_d$, the cooling to $T_s$, the unloading, and the recovery step to $T_d$ were performed consecutively with no (or minimal) intermediate holding times. By varying the deformation loads, the effect of increasing deformation strains ($\varepsilon_m^d$) on the instantaneous SM behavior was obtained. By increasing the number of successive SM cycles (N) at a given $\varepsilon_m^d$, the effect of thermo-mechanical aging on the instantaneous SM behavior was determined. Likewise, the influence of the heating recovery rate ($[dT/dt]_r$) on the instantaneous strain recovery was investigated. Finally, the effects of isothermally holding our SMP's at $T_s$ (fixing step) after unloading or $T_d$ (deformation step) after deforming were studied.

## DISCUSSION

The thermo-mechanical properties of our epoxy-SMP's (see Scheme 1) are summarized in Table I. The SM behavior of our epoxy-SMP's is discussed below in light of their molecular composition and physical properties relative to the varying experimental conditions.

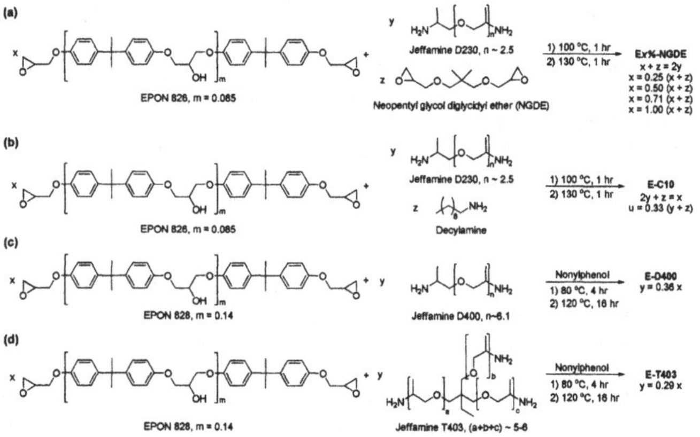

**Scheme 1.** Chemical composition and curing conditions of our epoxy-SMP's.

**Table I.** Thermo-mechanical properties of our epoxy SMP's.

|  | E-D400 | E-T403 | E-C10 | E25-NGDE | E50-NGDE | E75-NGDE | E100-NGDE |
|---|---|---|---|---|---|---|---|
| $T_g = T_{trans}$ (°C)* | 44.0 ± 0.6 | 75.3 ± 0.9 | 62.6 ± 0.7 | 31.4 ± 0.3 | 48.1 ± 0.7 | 72.0 ± 0.2 | 93.1 ± 1.9 |
| tan δ | 1.24 | 0.70 | 1.63 | 1.70 | 1.38 | 1.14 | 1.16 |
| $T_s$ (°C) | 10.4 ± 3.8 | 15.9 ± 1.9 | 33.1 ± 1.2 | 4.4 ± 3.1 | 7.5 ± 1.6 | 21.6 ± 5.9 | 31.8 ± 4.5 |
| $T_d$ (°C) | 87.1 ± 0.0 | 111.4 ± 0.2 | 101.2 ± 2.7 | 71.3 ± 1.2 | 93.6 ± 2.3 | 115.8 ± 3.2 | 130.9 ± 1.0 |
| $E'_d$ (MPa) | 9.25 ± 0.09 | 17.20 ± 0.28 | 6.02 ± 0.31 | 4.50 ± 0.9 | 9.46 ± 0.40 | 14.34 ± 0.42 | 18.89 ± 2.49 |
| $E'_d/E'_s$ | 0.0038 ± 0.0006 | 0.0057 ± 0.0002 | 0.0025 ± 0.0001 | 0.0023 ± 0.0002 | 0.0054 ± 0.0002 | 0.0050 ± 0.0002 | 0.0070 ± 0.0005 |
| $T_r$ ‡ | 38.7 | 76.8 | 60.4 | 27.0 | 50.1 | 71.2 | 92.1 |
| $v_e$ (x10⁻³ mol/g) | 2.12 | 2.57 | 3.24 | 4.03 | 3.60 | 3.25 | 2.97 |

*Measured by DMA from the apex of tan δ; ‡Calculated from the strain recovery response when heating at 3 °C/min.

## Effect of increasing number of shape memory cycles

The evolution of the deformation strain with increasing cycle numbers under constant deformation stress is shown in Figure 1(a) for all samples. The corresponding apparent storage modulus at $T_d$ as a function of cycle number ($E_d'^{app}(N)$) is shown in Figure 3(b). $E_d'^{app}(N)$ leveled off after the second cycle for all samples but E-C10. Indeed, E-C10 showed a continuous decrease of the strain with increasing N (see Figure 1(a)), alternatively a continuous increase of $E_d'^{app}(N)$ (see Figure 1(b)). We believe that this is due to a rearrangement of the pendant decyl chains in E-C10 along the deformation direction. Upon heating through the transformation (glass) transition, the *confined* decyl chains cannot completely recover their initial state and remain somewhat oriented along the deformation direction. Hence, the permanent network conformation is only partially recovered. A degree of molecular anisotropy is preserved causing property anisotropy (i.e., an increase of the modulus in the deformation direction).

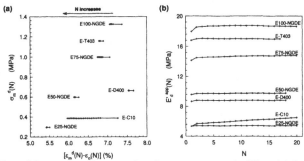

**Figure 1.** Effect of the number of consecutive shape memory cycle (N) on (a) the stress-strain relationship during the deformation stage, and (b) the apparent storage modulus at the deformation temperature for each cycle ($E_d'^{app}(N)$).

Except for the E-C10 sample, all samples reached stable values of shape fixity and shape recovery at $N \geq 2$. $R_f$ and $R_r^{min}$ for E-C10 strongly depended on $E_d'^{app}$ and the thermal contraction of the sample upon cooling, and decreased with N. In contrast, $R_r^{eu}$ was not affected by the change in $\varepsilon_m^d$ and $E_d'^{app}$, and its value remained constant at about 100% for all samples.

Neither the response temperature ($T_r$) nor the recovery speed ($V_r$) were significantly influenced by the succession of shape memory cycles.

### Effect of Varying the Deformation Load

As shown in Figure 2, the *instantaneous* shape memory behavior of our epoxy-SMP's appears independent of their structure, composition, crosslink density and modulus. Indeed, $R_f$, $R_r^{min}$ and $R_r^{eu}$ of all samples could each be fitted using a single exponential regression.

Also, it was found that $R_f$ and $R_r^{min}$ varied linearly with the thermal contraction of the samples upon cooling, which changed with the deformation strain as a result of molecular anisotropy. The effect of thermal contraction on the shape memory response of our epoxies became negligible compared to $\varepsilon_m^d$ at strains above 6 to 7% (i.e., $R_r^{min}$ and $R_f$ ~100%). $R_r^{eu}$ approached 100% at all strain levels because it is independent of CLTE and thermal contraction and because strain relaxation was negligible.

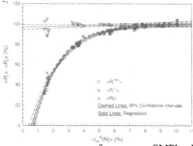

**Figure 2.** Instantaneous shape memory response of our epoxy-SMP's. Each point represents one epoxy at every $\varepsilon_m^d$.

As expected, for all samples the recovery speed ($V_r$) increased with strain and the recovery time ($t_r$) varied inversely, ranging from 0.58 to 3.19 min. Moreover, as shown in Figure 3, $V_r$ increased with increasing tan $\delta$, with decreasing $E'_d/E'_s$, and with increasing $v_e$. However, at lower crosslink density ($v_e$), the influence of chain mobility/flexibility (tan $\delta$) becomes the dominant factor governing the shape memory response.

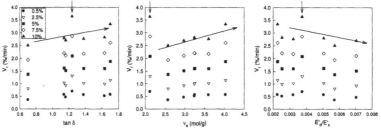

**Figure 3.** Variation of the recovery speed ($V_r$) as a function of (a) the loss angle (tan $\delta$), (b) the apparent crosslink density ($v_e$), and (c) $E'_d/E'_s$ of all epoxy-SMP's.

### Effect of Varying Isothermal Steps at $T_d$ and $T_s$

Depending on application requirements, an SMP may have to be held at high or low temperatures ($T>T_{trans}$ or $T<T_{trans}$) for extended periods of time prior to cooling and fixing or prior to recovery. Experimentally, this was mimicked by holding isothermally the SMP for 0, 15, 30, 60 and 180 min at $T_d$ after deformation or $T_s$ after unloading. Apart from E-C10, all samples showed good strain retention (i.e., negligible strain relaxation) even at the 180 min isothermal hold at $T_d$ or $T_s$ as shown in Figures 4 and 5, respectively.

At 180 min isothermal hold at $T_d$, E-C10 showed a reduction in shape recovery of 5.02 and 6.45% for $R_r^{min}$ and $R_r^{eu}$, respectively, and a slight increase in shape fixity of 1.40%. At 180 min isothermal hold at $T_s$, the E-C10 sample showed a slight reduction in $R_r^{min}$ and $R_f$ of 1.14 and 1.02%, respectively. In both cases, these effects resulted from the slight increase in $E'_d^{app}$ with N due to molecular rearrangement, which in turn caused a variation of the sample's thermal contraction on cooling. In addition, the longer residence time at $T_d$ also caused a reduction of

$R_r^{eu}$ for the E-C10 sample due to strain relaxation of the mobile decyl chains to adopt a more energetically favorable (coiled) conformation. This plastic deformation disabled the material to recover its initial shape when heated back to $T_d$ under no load, thereby reducing the value of $R_r^{eu}$.

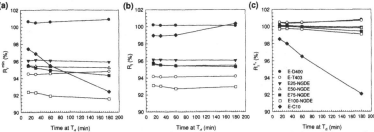

**Figure 4.** Effect of increasing the isothermal step at $T_d$ under load on (a) the minimum shape recovery ($R_r^{min}$), (b) the shape fixity ($R_f$), and (c) the shape recovery ($R_r^{eu}$) for all our epoxies.

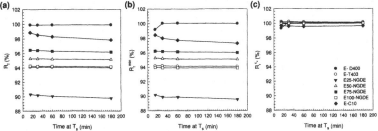

**Figure 5.** Effect of increasing the isothermal step at $T_s$ under no load on (a) the shape fixity ($R_f$), (b) the minimum shape recovery ($R_r^{min}$), and (c) the shape recovery ($R_r^{eu}$) for all our epoxies.

## Effect of Varying the Recovery Step Heating Rate

While higher recovery heating rates led to faster recovery speed (i.e., shorter recovery times ($t_r$)), it also led to strain recovery spanning over wider temperature windows and delayed recovery completion. At 17.28 °C/min 90% strain is recovered after 3.5 min, whereas, 26.8 min is necessary to achieve the same 90% strain recovery at 1 °C/min. Moreover, about 90% of the strain is recovered at 60 °C by heating at 1 °C/min but only at 87 °C by heating at 17.28 °C/min. This is most likely due to the low thermal conductivity of the samples. Above 10 °C/min, $T_r$ and $t_r$ leveled off indicating that the heat transfer process became the limiting factor for the SM response under our specific experimental setup (i.e., sample geometry, air environment).

## Cycle Time

The total cycle time ($t_{sm}$) of our epoxies ranges from 45.5 to 64.5 min. These values were not optimized and depended largely on the experimental conditions used. By optimizing the experimental conditions, $t_{sm}$ could potentially be reduced by half or more for all epoxies. In general, we found that shorter $t_{sm}$ can be expected from materials with higher tan $\delta$ and lower $E'_d/E'_s$.

## CONCLUSIONS

The shape memory behavior of epoxy-amine SMP's was systematically studied under varying experimental conditions. The results were discussed in terms of the materials structural and physical properties. It was found that the *instantaneous* response of our epoxy-SMP's was independent on composition, structure, and mechanical properties but was dominated by the coefficient of linear thermal expansion. The latter was found to vary with the deformation strains owing to molecular anisotropization of the deformed network. Above 6 to 7% strain, $R_f \approx R_r^{min} \approx R_r^{eu} \approx 100\%$.

More demanding experimental conditions caused a reduction in the SM properties of the networks with a lower crosslink density and/or a higher chain flexibility/mobility as a result of molecular rearrangement leading to property anisotropy and/or strain relaxation.

In general, the recovery speed ($V_r$) increased with increasing crosslink density ($v_e$), decreasing $E'_d/E'_s$, and increasing tan $\delta$ at all strain levels. However, at low crosslink density, the effect of tan $\delta$ becomes dominant. Expectedly, the recovery time varied inversely to $V_r$.

Although material improvement and thermo-cycling conditions optimization could greatly reduced the overall cycle time ($t_{sm}$); it was also found that $t_{sm}$ decreased with increasing tan $\delta$ and decreasing $E'_d/E'_s$.

In general, weakly crosslinked epoxy and epoxy containing highly flexible pendant chains should be avoided for high performance applications due to their diminished SM properties under more demanding applications even though they may show faster instantaneous response as does E-D400.

## REFERENCES

1. Y.Y.F.C. Vili, *Textile Research Journal* **77**(5), 290 (2007).
2. B, Dietsch, T. Tong, *Journal of Advanced Materials* **39**(2), 3 (2007).
3. D. Ratna, J. Karger-Kocsis, *Journal of Materials Science* **43**(1), 254 (2008).
4. G.P. McKnight, C. P. Henry, *Proceedings of SPIE* **6929** (*Behavior and Mechanics of Multifunctional and Composite Materials*), 692919/1 (2008).
5. W. Sokolowski, A. Metcalfe, A. Hayashi, L.H. Yahia, J. Raymond. J., *Biomedical Materials* **2**(1), S23 (2007).
6. T. Xie, I.A. Rousseau, *Polymer* **50**(8), 1852 (2009).
7. N.G. Sahoo, Y.C. Jung, N.S. Goo, J.W. Cho, *Macromolecular Materials and Engineering* **290**(11), 1049 (2005).
8. M. Y. Razzaq, L. Frormann, *Polymer Composites* **28**(3), 287 (2007).
9. N.G. Sahoo, Y.C. Jung, J.W. Cho, *Materials and Manufacturing Processes* **22**(4), 419 (2007).
10. T. Ohki, Q.-Q. Ni, N. Ohsako, M. Iwamoto, *Composites, Part A Applied Science and Manufacturing* **35**A(9), 1065 (2004).
11. I.A. Rousseau, *Polymer Engineering and Science* **48**(11), 2075 (2008).
12. I.A. Rousseau, T. Xie, *Proceedings of SPIE* (2009) (in Press).

Mater. Res. Soc. Symp. Proc. Vol. 1190 © 2009 Materials Research Society          1190-NN01-09

# Ability to control the glass transition temperature of amorphous shape-memory polyesterurethane networks by varying prepolymers in molecular mass as well as in type and content of incorporated comonomers

J. Zotzmann[1], S. Kelch[2], A. Alteheld[3], M. Behl[1] and A. Lendlein[1]
[1]Center for Biomaterial Development, GKSS Research Center Geesthacht GmbH,
  D-14513 Teltow, Germany
[2]Sika Technology AG, Tüffenwies 16, CH-8048 Zurich, Switzerland and
[3]BASF Aktiengesellschaft, Carl-Bosch-Str. 38, D-67056 Ludwigshafen, Germany

## ABSTRACT

The need of intelligent implant materials for applications in the area of minimally invasive surgery leads to tremendous attention for polymers which combine degradability and shape-memory capability. Application of heat, and thereby exceeding a certain switching temperature $T_{sw}$, causes the device to changes its shape. The precise control of $T_{sw}$ is particularly challenging. It was investigated how far the glass transition temperature $T_g$ of amorphous polymer networks based on star-shaped polyester macrotetrols crosslinked with a low-molecular weight linker can be controlled systematically by incorporation of different comonomers. The molecular weight of the prepolymers as well as type and content of the comonomers was varied. The $T_g$ could be adjusted by selection of comonomer type and ratio without affecting the advantageous elastic properties of the polymer networks.

## INTRODUCTION

Aliphatic (co)polyesters are used today in numerous medical applications such as degradable polymeric implant materials for surgical devices and drug release systems.[1-3] The combination of degradability with the shape-memory capability resulting in multifunctional polymers is motivated by the aim to insert spacious degradable implants through small incisions into the body[4] or to manipulate their shape on demand during a minimally-invasive procedure.[5,6] The shape-memory effect enables a bulky device to be programmed into a compressed temporary shape. In the thermally induced shape-memory effect, at $T_{sw}$ the device recovers its original shape. According to various application strategies, $T_{sw}$ must be adjustable in different temperature ranges. If $T_{sw}$ ranges between room temperature and body temperature the thermally-induced shape change occurs automatically after implantation. However, a $T_{sw}$ slightly above body temperature enables on demand control of the shape change. Here, the heat is applied either indirectly through IR-irradiation or directly by application of an external heating medium. $T_{sw}$ of amorphous polymers is related to $T_g$ of the switching domain. Previous investigations showed the possibility of controlling $T_{sw}$ and the hydrolytic degradation rate of a shape-memory polymer network with crystallizable switching segments where $T_{sw}$ was related to the melting point $T_m$ of the switching phase.[7] Alteration of the macromolecular architecture to completely amorphous polymer networks allows a broadened adjustability of morphology and mechanical properties. Unlike semi-crystalline polymers these materials show a more homogeneous degradation being advantageous for implanted medical matrices.
It is beneficial for biomaterials being applied in implants if their mechanical properties can be adjusted to approximate those of the target tissue. In our investigations a polymer architecture

based on star-shaped precursors was selected because of the advantageous elastic properties of such networks.[5,8] Star-shaped polyester macrotetrols were crosslinked with a diisocyanate resulting in polyesterurethane networks (see Figure 1).[9] We investigated if $T_g$ of these amorphous, degradable shape-memory polymer networks can be controlled by incorporation of different comonomers. *rac*-Dilactide was chosen as basic monomer since its amorphous (co)polymers have been proven to be biodegradable and are widely used for medical applications. However, the $T_g$ of about 59 °C is too high to be used as $T_{sw}$ in the temperature range between room temperature and slightly above body temperature. Diglycolide, ε-caprolactone and p-dioxanone were used as comonomers in the prepolymer generating ring-opening polymerization, whose homopolymers have lower $T_g$s than poly(*rac*-dilactide). Two parameters were evaluated in their ability of controlling $T_g$. First, the influence of the molecular weight of the star-shaped prepolymers on $T_g$ of the networks was investigated. Second, the possibility to adjust $T_g$ of the networks by altering the comonomer content of the prepolymers was determined. The strategy to adjust the $T_g$ of the copolymer networks is derived from the Fox relation, which allows estimating $T_g$ of amorphous copolymers from $T_g$s of the related homopolymers formed by each of the comonomers.[10]

## EXPERIMENT

Macrotetrols were prepared by ring-opening copolymerization of lactones using pentaerythrite as initiator. The crosslinking of the prepolymers was obtained by polyaddition with diisocyanate resulting in copolyesterurethane networks (Figure 1). $M_n$ was measured by GPC in chloroform, comonomer contents were calculated from NMR spectra. Thermal transitions were recorded by DSC measurements using a Perkin Elmer DSC7 apparatus. Mechanical properties were determined on a ZWICK1425 tensile tester at 37 °C. Details of material synthesis and characterization methods are described in reference [9].

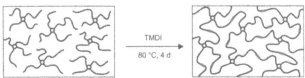

**Figure 1.** Scheme of network synthesis from star-shaped precursors, open circles: netpoints, black lines: polyester chains, grey rectangles: diurethane links, TMDI: isomeric mixture of 1,6-diisocyanato-2,2,4-trimethylhexane and 1,6-diisocyanato-2,4,4-trimethylhexane.

## DISCUSSION

### Influence of the molecular mass on $T_g$

According to DSC measurements, all synthesized macrotetrols were amorphous since all investigated prepolymers showed only a single thermal transition characterized by a $T_g$. The influence of the molecular weight on $T_g$ of the prepolymers can be evaluated with the Fox-Flory theory which describes the $T_g$ as non-linear function of the molecular mass.[11] The observed dependence of $T_g$ of the macrotetrols from $M_n$ is shown in Figure 2. Star-shaped prepolymers with a constant comonomer content of 17 wt% glycolide, 16 wt% ε-caprolactone, and 13 wt% p-dioxanone respectively were investigated.

The values for $T_g$ increased as expected with increasing $M_n$ of the synthesized prepolymers having $M_n$ values in a range from 1000 g·mol⁻¹ to 12000 g·mol⁻¹. The given regression functions according to the Fox-Flory theory show a good correlation. $T_g$ became nearly independent from $M_n$ when $M_n$ was increased above 10000 g·mol⁻¹ as the influence of the end-groups decreases with increasing chain length.

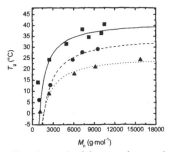

**Figure 2.** $T_g$ (from DSC second heating run) of the prepolymers depending on $M_n$ and regression functions according to the Fox-Flory theory; squares: P-LG prepolymers, triangles: P-LC prepolymers, circles: P-LD prepolymers. P: pentaerythrite initiator, L: *rac*-dilactide monomer, G: diglycolide comonomer, C: ε-caprolactone comonomer, D: p-dioxanone comonomer.

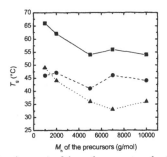

**Figure 3.** $T_g$ (DSC second heating run) of the polymer networks depending on $M_n$ of the precursors with a constant comonomer content; squares: N-P-LG(17) networks, triangles: N-P-LC(16) networks, circles: N-P-LD(13) networks; lines: guide for the eye.

The dependence of the $T_g$ of the polymer networks from the molecular weight of the prepolymers is depicted in Figure 3. The $T_g$s of the polymer networks were 10 to 15 °C higher than those of the prepolymers due to the limited mobility of the polymer chains within the networks. The $T_g$ decreasing effect of shorter chain lengths which was observed in the prepolymers becomes negligible in the networks because the end-groups are linked to the network. The polymer networks from low-molecular weight prepolymers have a higher crosslinking density and therefore show a slight dependence from $M_n$ of the prepolymers. However, the effect was weak and ceased for networks from prepolymers with $M_n$ higher than 4000 g·mol⁻¹. The N-P-LC networks showed no consistent dependence from $M_n$ of the P-LC at all. In this context, the molecular mass of the prepolymers was found to be no suitable parameter to conveniently control the $T_g$ of the polymer networks.

**Influence of comonomer content on $T_g$**

It was investigated if the comonomer content of the prepolymers is a suitable parameter for $T_g$ adjustment in the polymer networks. The comonomer content was varied while $M_n$ of the star-shaped prepolymers was kept constant at around 10000 g·mol$^{-1}$. Thus, the prepolymers were in the $M_n$-range where $T_g$ is independent from $M_n$. Furthermore, this relatively high $M_n$ of the prepolymers resulted in polymer networks having a high length of chain segments and enabled good elastic properties. $T_g$ of the prepolymers decreased with increasing comonomer content (Figure 4) and was in accordance with the Fox relation for amorphous copolymers. The content of the comonomer was varied between 0 and 52 wt% for diglycolide, 53 wt% for the ε-capro-lactone and 65 wt% for p-dioxanone respectively. Within that range $T_g$ of the P-LG prepolymers decreased by 5 K. The P-LC and P-LD prepolymers displayed a stronger influence of comono-mer content on $T_g$ resulting in a $T_g$ decrease of 62 and 70 K respectively. This behavior is attri-buted to the lower $T_g$s of the corresponding homopolymers of these comonomers.

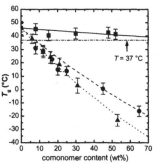

**Figure 4.**[12] $T_g$ (from DSC second heating run) of the prepolymers depending on comonomer content and regression function. The bars show the temperature interval of the glass transition; squares: P-LG prepolymers, triangles: P-LC prepolymers, circles: P-LD prepolymers. The bars show the temperature interval of the glass transition.

The same dependence of $T_g$ was also observed for the polymer networks (see Figure 5). In all three series the $T_g$ decreased with increasing comonomer content. N-P-LG networks with $T_g$s slightly above body temperature could be obtained. The observed temperature range of $T_g$ of the N-P-LC and the N-P-LD network series covered a very broad range between -23 and 45 °C. The $T_g$s of these polymer networks with comonomer contents between 5 and 20 wt% were adjustable in a temperature range between room temperature and body temperature.

**Mechanical Properties**

The network architecture based on star-shaped precursors allows the control of elasticity by altering the chain length between the crosslinks which is defined by the molecular weight of the precursors. Longer chains and the resulting lower crosslink density lead to a higher elasticity of the polymer network in the rubbery elastic state without yielding and with higher values of the strain at break ($\varepsilon_B$). In this context it was investigated if type and content of the comonomers influence the mechanical properties in a way that deteriorates the advantage of the polymer network architecture. It was observed that the elasticity modulus ($E$) was not influenced by the comonomer content. However, a strong dependence of the elasticity of the N-P-LC and N-P-LD series from the thermal properties was obtained. (Figure 6a) $E$ values remained high for the N-P-LG networks which were all glassy at 37 °C. The $T_g$ for the N-P-LC and N-P-LD networks decreases below body temperature for comonomer contents higher than 15 wt%. For this reason the polymer networks of this composition were soft at 37 °C resulting in substantially lower $E$ values.

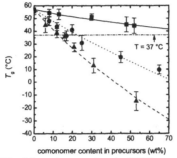

**Figure 5.**[12] Dependence of $T_g$ of the polymer networks (DSC second heating run) on comonomer content of prepolymers used as precursors in wt%. squares: N-P-LG networks, triangles: N-P-LC networks, circles: N-P-LD networks. The curves were calculated according to the Fox relation. The bars show the temperature interval of the glass transition.

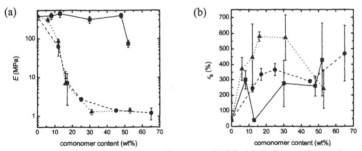

**Figure 6.** Mechanical properties from tensile tests at 37 °C: (**a**) $E$, (**b**) $\varepsilon_B$ of the polymer networks in dependence of the comonomer content; squares: N-P-LG networks, triangles: N-P-LC networks; circles: N-P-LD networks; lines: guide for the eye; the bars show the standard deviation around the average value of at least three tested samples.

The $\varepsilon_B$ values remained above 200% for a wide range of the comonomer content regardless of the type of comonomer (Figure 6b). The partially poor reproducibility of the measurements resulting in large errors might be attributed to the small number of the test specimens for statistics. Variation of the molecular weight of the prepolymers did not influence the $T_g$ of the

41

polymer networks. However, it is a parameter to alter the elastic properties of polymer networks from star-shaped precursors by determining the crosslink density. The concept of the $T_g$ being slightly higher than body temperature enables the application of such materials for implants which need a higher mechanical stress resistance. The materials become glassy when cooled to body temperature after triggering the shape-memory effect.

## CONCLUSIONS

$T_g$ of amorphous *rac*-lactide based shape-memory copolymer networks could be varied in a wide temperature interval by changing type and content of comonomers of star-shaped prepolymers. It was found that the molecular mass of the star-shaped copolyester prepolymers was no suitable parameter for the adjustment of the $T_g$ of the polymer networks. Copolyester-urethane networks from prepolymers with > 50 wt% diglycolide, 10 – 25 wt% ε-caprolactone or 15 – 45 wt% p-dioxanone showed $T_g$s in a range between room temperature and slightly above body temperature. Hence, $T_g$ can be controlled precisely by type and content of comonomer in the prepolymers. $T_g$ can be used as trigger for the thermally-induced shape-memory effect of amorphous shape-memory polymers. Therefore, the investigated polymer networks are expected to exhibit the same controllability of their $T_{sw}$. The mechanical properties were nearly indepen-dent from the chemical composition of the networks. We explored that $T_g$ can be varied while pertaining the network's molecular architecture, the crosslink density and consequently the good elastic properties. These multifunctional amorphous polymer networks hold great promise to be applied in regenerative therapies such as bioactive implants which require biodegradability and shape-memory properties. They represent a new group of materials exhibiting a $T_{sw}$ being controllable in a temperature interval that meets the requirements of various medical application strategies.

## REFERENCES

1    R. Langer and D. A. Tirrell, *Nature* **428**, 487 (2004).
2    M. Hakkarainen, A. Hoglund, K. Odelius, and A. C. Albertsson, *Journal of the American Chemical Society* **129**, 6308 (2007).
3    M. Vert, *Biomacromolecules* **6**, 538 (2005).
4    X. L. Lu, W. Cai, Z. Gao, and W. J. Tang, *Polymer Bulletin* **58**, 381 (2007).
5    C. C. Min, W. J. Cui, J. Z. Bei, and S. G. Wang, *Polymers for Advanced Technologies* **16**, 608 (2005).
6    M. C. Chen, H. W. Tsai, Y. Chang, W. Y. Lai, F. L. Mi, C. T. Liu, H. S. Wong, and H. W. Sung, *Biomacromolecules* **8**, 2774 (2007).
7    S. Kelch, S. Steuer, A. M. Schmidt, and A. Lendlein, *Biomacromolecules* **8**, 1018 (2007).
8    A. Alteheld, Y. Feng, S. Kelch, and A. Lendlein, *Angewandte Chemie-International Edition* **44**, 1188 (2005).
9    A. Lendlein, J. Zotzmann, Y. Feng, A. Alteheld, S. Kelch, *Biomacromolecules* **10**, (2009), accepted.
10   T. G. Fox, *Bulletin of the American Physical Society* **1**, 123 (1956).
11   T. G. Fox and P. J. Flory, *Journal of Applied Physics* **21**, 581 (1950).
12   Reprinted with permission from reference [9]. Copyright 2009 American Chemical Society.

Mater. Res. Soc. Symp. Proc. Vol. 1190 © 2009 Materials Research Society     1190-NN03-02

## Functional Fatigue of Shape-Memory Polymers

Christina Schmidt, Klaus Neuking, Gunther Eggeler
Ruhr-University Bochum, Institute for Materials, 44780 Bochum, Germany

## ABSTRACT

The present study represents a first step towards an understanding of what we refer to as the functional fatigue behaviour of shape-memory polymers. These materials have a processing shape B and a programmed shape A [1]. And when the material is exposed to an appropriate stimulus (in our case: heating above a critical temperature), a one way effect is observed: A → B (one way effect: 1WE). The objectives of the present study were to find out whether and how often programming can be repeated, whether repeated programming affects the 1WE and how much irreversible strain the material accumulates. We study the effect in dependence of different stress levels, and consider the effect of recovery temperature and recovery time. As a model material we examine the commercial amorphous shape-memory polymer Tecoflex® and subject it to 50 programming/1WE cycles. It turns out that programming, cooling, unloading and heating to trigger the 1WE causes an increase of irreversible strain and is associated with a corresponding decrease of the intensity of the 1WE in particular during the first thermomechanical cycles.

## INTRODUCTION

Shape-memory polymers (SMP) are polymers that are able to respond to external stimuli. SMP show a one way effect like shape-memory alloys. They can be deformed from a permanent shape to a temporary shape and recover by means of an external stimulus. For this direct thermal activation is the commonly used [2]. In contrast, alternative activation stimuli like inductive heating [3, 4], irradiation [5], magnetic [6] or electric fields [7] and pH changes [8] are not frequently applied. The shape-memory effect of SMP is based on the physical properties of the constituents of the polymer structure, chemistry and morphology.

Thermoplastic SMP are typically (phase segregated) multiblock copolymers which consist of at least two types of segments [9-11]. The component with the higher thermal transition $T_{hard}$, is referred to as hard segment. This hard segment (an elastomer) acts as physical netpoint and represents the part of the network which provides the mechanical strength of the material at temperatures above the transition temperature $T_{trans}$ of the SMP ($T_{trans} < T > T_{hard}$). The soft segment (a thermoplastic), having a lower thermal transition $T_{trans}$, stabilizes the hard segment at low temperatures ($T < T_{trans}$) and looses its strength for $T > T_{trans}$. The transition temperature $T_{trans}$ of the SMP can be a glass transition or a melting point. The shape-memory effect occurs by exposing the SMP to a specific switching temperature, $T_{sw}$. Compared to $T_{trans}$, the switching temperature $T_{sw}$ is result of a thermomechanical test [12,13].

In its processed shape B, the chains of the shape memory polymer are randomly oriented. Below $T_{sw}$, the copolymer shows only little elasticity (high Young's modulus). In contrast, above $T_{sw}$, the soft segment looses its stiffness (low Young's modulus). Above $T_{sw}$ the hard segment chains can be stretched from their random (high entropy) into an aligned (low entropy) configuration by mechanical deformation. Subsequent cooling stiffens the strained polymer chains and by this the low entropy stage is frozen in. The shape-memory polymer is now in its programmed shape A.

This shape A can be sustained indefinitely as long as $T < T_{sw}$. By heating the unconstrained SMP above $T_{sw}$ the initial shape is re-established because the soft segments no longer stabilize the low entropy configuration. For $T > T_{sw}$ entropic forces act as the driving force for the recovery. There is a good understanding of the physical mechanisms which are responsible for the one way effect in SMPs and potential applications have been identified. But there is only limited information available on how often programming/1WE-cycles can be repeated and how this affects SMP properties. In the present study this type of functional fatigue in the commercial SMP Tecoflex[®] subjected to 50 programming/1WE cycles has been examined.

**EXPERIMENTS**

For our tensile tests we used SMP strips of the totally amorphous shape-memory thermoplastic elastomer Tecoflex[®] EG72D (TFX) (Lubrizol, USA). The strip dimensions were $1.5 \times 6 \times 50$ mm[3]. TFX is a cycloaliphatic polyetherurethane which is synthesized from methylene bis(p-cyclohexyl isocyanate) (H12MDI), 1,4-butanediol (1,4-BD) and poly(tetramethylene glycol) (PTMG) [14, 15]. This multiblock copolymer contains H12MDI/1,4-BD hard segments and H12MDI/PTMG soft segments. It additionally forms a mixed phase, which is the switching phase. The transition temperature of the switching phase is a glass transition and occurs at $T_{trans} = 74°C$. Tecoflex[®] EG72D was purchased as granulate for injection moulding. Our injection moulding parameters are listed in Table I. For injection moulding we use an Arburg Allrounder 270 M 500-210 extruder.

**Table I.** Injection moulding parameters for Tecoflex[®] EG72D.

| Melt temperature of TFX (°C) | 270 |
|---|---|
| Mould temperature (°C) | 60 |
| Injection rate (mm/s) | 26 |
| Injection pressure (MPa) | 60 |
| Holding pressure (MPa) | 55 |
| Holding time (s) | 15 |

Tecoflex[®] samples were tested using a Zwick Z2.5 (Ulm, Germany) test rig with a testXpert[®] control and data acquisition system equipped with a temperature chamber to adjust temperatures. 1000 N respectively 100 N load cells were used for ambient ($T_{amb} = 21°C$) and high temperature ($T_{high} = 75, 80, 85, 90°C$) testing. Several tests at $T_{high}$ were performed to characterize the temperature dependence of the mechanical properties above $T_{sw}$. Tests were performed at crosshead speeds of 5 mm/minute. Strains were derived from crosshead displacements. Heating/cooling rates of 5 K/minute were applied between 90 respectively 80 and 15°C in the programming/1WE cycles which were performed for maximum technical strains of 100, 200, 400 and 600%. The time interval for completion of the 1WE on heating was 10 minutes, before the next cycle was imposed. For each cycle we determine a recovery ratio which is given by

$$R = (\varepsilon_m - \varepsilon_{irr,N})/\varepsilon_m \qquad (1)$$

where $\varepsilon_m$ is the maximum (programming) technical strain and $\varepsilon_{irr,N}$ is the irreversible technical strain which the one way effect cannot recover. All experiments were taken through to 50 cycles where cycling was interrupted.

Dynamic mechanical analysis (DMA) using pull-pull loading was used to measure the glass transition temperatures $T_g$ using an Eplexor 500 N (Gabo Qualimeter). The samples were heated from -50 to 150°C at a heating rate of 2 K/minute. Testing was performed under cyclic tensile strain control imposing a mean strain of 0.25% and superimposing a sinusoidal cyclic strain of ± 0.1% at 10 Hz.

## RESULTS AND DISCUSSION

### Dynamic Mechanical Analysis (DMA)

Our DMA results are presented in Figure 1, where the stored modulus (frequency dependent elastic modulus) and the loss factor tan δ (mechanical damping) are presented as a function of temperature. The peaks of the loss factor vs. temperature curve indicate the temperatures where maximum damping is observed. The first weak maximum around -40°C corresponds to the glass transition temperperature of the H12MDI/PTMG soft domains [16]. The pronounced peak maximum temperature at 74°C reflects the glass transition temperature $T_g$ (= $T_{trans}$) of the mixed phase and the third peak at 120°C is related to the softening temperature of the hard domains [17].

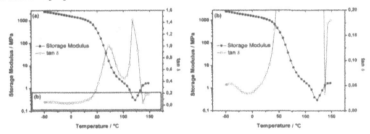

**Figure 1.** Storage modulus (full symbols) and damping as a function of temperature. The maxima of the loss factor are due to the glass transitions of the soft segments (expanded in (b): $T_g$ = -40°C), the mixed phase ($T_g$ = $T_{trans}$ = 74°C) and the softening temperature of the hard segments ($T_s$ = 120°C).

### Tensile Testing

The stress strain curves are presented in Figures 2 (a) and (b).

45

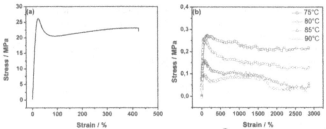

**Figure 2.** Stress strain curves from tensile tests on Tecoflex®. (a) Ambient temperature: 21°C. (b) T > $T_{sw}$: 75, 80, 85, 90°C.

At room temperature, there is a linear increase of stress up to 26 MPa at 24 %, from where it drops to a plateau stress of 22 MPa. Rupture occurs at a technical strain of 425 %, Figure 2(a). At temperatures above the transition temperature $T_{trans}$ a completely different behaviour is observed. Technical strain at failure reaches more than 2000%. Maximum stresses depend on the test temperature and reach no more than 0.1 to 0.3 MPa. With increasing temperature the material softens. In our experiments no ruptures occurred. Instead the material plastically flows without breaking.

**Programming and 1WE**

Figure 3 exemplary shows the thermomechanical behaviour of Tecoflex®. Starting with programming and subsequently shape resetting by the one way effect (1WE), repeated cycles at 80 and 90°C with technical strains up to 100 and 400% and recovery time of 10 minutes are shown.

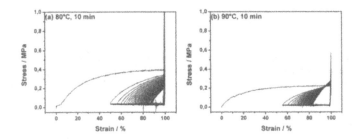

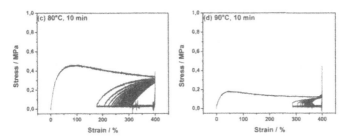

**Figure 3.** Shape-memory cycles at 10 minutes recovery time: (a,b) 100%, 80°C / 90°C, (c,d) 400%, 80°C / 90°C.

The results show how technical strains evolve with time during cyclic thermomechanical loading. It can be seen that the stresses and width of hysteresis at a testing temperature of 80°C are larger than at 90°C. At 80°C the maximum stresses reach values of 0.4 MPa while at 90°C maximum stresses correspond approximately to 0.2 MPa at higher irreversible strains.

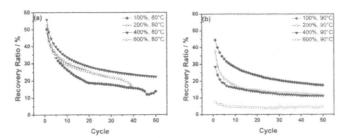

**Figure 4.** Recovery ratio as a function of the number of full shape-memory cycles: (a) 80°C testing temperature, (b) 90°C testing temperature.

In Figure 4 we show the evolution of the recovery ratio as a function of the number of cycles. It can be seen that at 80°C higher recovery ratios are obtained for higher maximum strains, while at 90°C a reverse effect can be observed. Within the first few cycles the amount of irreversible strain increases and accumulates to a limit, which depends on the maximum strain. The recovery ratio for the programming cycles at 90°C is generally lower than at 80°C.

## SUMMARY AND CONCLUSION

In the present study we performed thermomechanical experiments on the shape-memory polymer Tecoflex®. The following results were obtained and conclusions can be drawn:
(1) Dynamic mechanical analysis showed maximum damping at the glass temperature $T_g$ of the soft segments at -40°C, at the glass temperature $T_g$ of the mixed phase at 74°C and at the softening temperature $T_s$ of the hard segments at 120°C.
(2) Tensile tests were conducted at 21, 75, 80, 85 and 90°C. At 21°C, below the switching temperature, the yield stress is approx. 26 MPa and technical strain at failure approx. 425 %. At

temperatures above the switching temperature, the material can be strained up to more than 2000 % without rupture while maximum stresses are in the range of 0.1 to 0.3 MPa. With an increase in temperature the maximum stresses decrease and the material softens.

(3) Thermomechanical cycles including programming, cooling, unloading and heating to trigger the 1WE were examined. Cycles were performed at 80 and 90°C with maximum technical strains of 100, 200, 400 and 600 % and a recovery time of 10 minutes. It was found that higher maximum technical strains $\varepsilon_m$ result in higher recovery ratios at 80°C, while at 90°C a reverse effect is observed. Recovery ratios at 80°C are also generally higher than at 90°C. During the first cycles the accumulation of irreversible strain saturates out. This is associated with plastic deformation of the polymer, which is probably due to the displacements of elongated polymer chains relative to each other.

## ACKNOWLEDGMENTS

The authors acknowledge funding through project B3 of the collaborative research centre SFB459 (shape-memory technology) funded by the Deutsche Forschungsgemeinschaft (DFG) and the State North Rhine Westphalia (NRW).

## REFERENCES

1. M. Behl and A. Lendlein, *materials today* **10**, 20 (2007).
2. M. Behl and A. Lendlein, *Soft Matter* **3**, 58 (2007).
3. H. Koerne, G. Price, N. Pearce, M. Alexander and R. A. Vaia, *Nat. Mater.* **2**, 115(2004).
4. P. R. Buckley, G. H. McKinley, T. S. Wilson, W. Small IV, W. J. Benett, J. P. Bearinger, M. W. McElfresh and D. J. Maitland, *IEEE Trans. Biomed. Eng.* **53**, 2075 (2006).
5. S. V. Ahir and E. M. Terentjev, *Nat. Mater.* **6**, 491 (2005).
6. Y. Wang, Y. Hu, X. Gong, W. Jiang, P. Zhang and Z. Chen, *J. Appl. Polym. Sci.* **103**, 3143 (2007).
7. T. J. Luand and A. G. Evans, *Sens. Actuators A* **99**, 290 (2002).
8. K. D. Harris, C. W. M. Bastiaansen and D. J. Broer, *J. Microelectromech. Syst.* **16**, 480 (2007).
9. J. Hu, in *Shape Memory Polym. and Textiles* (Woodhead Publishing Limited and CRC Press LLC), (2007).
10. K. Neuking, A. Abu-Zarifa, S. Youcheu-Kemtchou and G. Eggeler, *Adv. Eng. Mat.* **7**, 1014 (2005).
11. A. Lendlein and S. Kelch, *Angew. Chem.* **114**, 2138 (2002).
12. A. J. Coury, P. C. Slaikeu, P. T. Cahalan, K. B. Stokes, C. M. Hobot, *J. Biomater. Appl.* **3**, 130 (1988)
13. P. Miaudet, A. Derré, M. Maugey, C. Zakri, P. M. Piccione, R. Inoubli, P. Poulin, *Science* **318**, 1294 (2007)
14. R. Mohr, K. Kratz, T. Weigel, M. Lucka-Gabor, M. Moneke, A. Lendlein, *PNAS* **103**, 3540 (2006).
15. A. Reddy, E. Arzt, A. del Campo, *Adv. Mater.* **19**, 3833 (2007).
16. R. E. Solis-Correa et al., *J. Biomater. Sci. Polymer Edn.* **18**, 561 (2007).
17. N. M. K. Lamba, K. A. Woodhouse, S. L. Cooper, in *Polyurethanes in Biomedical Applications* (CRC Press), (1998).

Mater. Res. Soc. Symp. Proc. Vol. 1190 © 2009 Materials Research Society   1190-NN03-18

# X-ray scattering studies to investigate triple-shape capability of polymer networks based on poly(ε-caprolactone) and poly(cyclohexyl methacrylate) segments

Wolfgang Wagermaier, Dieter Hofmann, Karl Kratz, Marc Behl and Andreas Lendlein
*Center for Biomaterial Development, GKSS Research Center Geesthacht GmbH,
Kantstr. 55, 14513 Teltow, Germany*

## ABSTRACT

The super-molecular structure and morphology of shape-memory polymers (SMP) have an evident influence on the shape-memory effect (SME). More detailed information on these structure-function relations during the dynamic processes of programming and shape recovery are required to better understand the SME.

Here we explore whether wide and small angle x-ray scattering (WAXS, SAXS) in combination with deformation experiments can help to characterize and better understand the respective materials super-molecular structure (spatial organization of chain segments in crystalline and non-crystalline regions, characterized by parameters such as crystallinity, crystallite-sizes, domain-sizes and -arrangements) and its changes upon varying mechanical loads and temperature increase as stimulus. Multiphase polymer networks based on poly(ε-caprolactone) and poly(cyclohexyl methacrylate), whose molecular structures allow formation of at least two separated domains, were investigated using WAXS and SAXS, to describe the respective super-molecular structures and morphologies and their development during cyclic, thermomechanical tensile tests reproducing key features of shape-memory programming and recovery.

The creation of the triple-shape capability for this AB polymer network system is performed by a one-step process, which is similar to a conventional dual-shape programming process. It could be shown via SAXS that a long period between crystalline domains exists for these polymer networks. The value of this long period changes by some nanometers as a consequence of programming and the resulting elongation of the respective sample. Further insights could be obtained by investigating WAXS diffraction peaks, detected at different steps during the thermomechanical treatment. It could be shown that crystal sizes in this polymer system remain unaffected by the programming process, while the crystallization of the stretched samples during the cooling process leads to a spatial rearrangement (preferential orientation) of crystalline domains.

## INTRODUCTION

SMP can move from a first shape (A) to a second shape (B), whereas the shape change is induced by a stimulus such as heat or light [1, 2]. In addition to these so called dual-shape polymers recently triple-shape polymers were introduced, which can move to a third shape (C) [3]. This triple-shape capability is obtained for multiphase polymer networks after application of a specific thermomechanical programming process consisting of two steps. Recently it could be demonstrated for an AB polymer network system that a triple-shape capability could be created by a simple one-step process, which is similar to a conventional dual-shape programming process [4]. This finding is remarkable, as a similar one-step process resulted only in a dual-shape capability for grafted polymer networks [5].

The SME results from the polymer architecture / morphology and the applied shape-memory creation procedure (SMCP), which is often named programming. Programming is followed by a recovery module, which can either be applied under stress-free or constant strain conditions. Programming and recovery are complex processes, which are influenced by many parameters (temperature, degree and rate of stretching, type of mechanical loading). The SME is investigated on a macroscopic level by cyclic, thermomechanical tests. The underlying mechanism of the SME, based on changes on the morphological level needs to be further investigated.

X-ray scattering is an appropriate method to investigate the morphology of polymers, especially if semi-crystalline components are involved. In this way, structure information of polymers from different length scales can be determined. Wide angle X-ray scattering (WAXS) patterns give information on the spatial arrangement of polymer chain segments (e.g. crystallinity, crystalline structure, size of crystals, crystal distortions and possible preferential orientations of chain segments in non-crystalline regions). In the small-angle X-ray scattering (SAXS) range typical nanostructure features (e.g. domain sizes and long periods between crystalline domains in semi-crystalline materials) can be observed. By combining tensile testing and X-ray scattering techniques structure and deformation mechanisms in materials can be well explained [6,7].

SMP with dual-shape capability consist of at least one phase, which can form physical crosslinks. These are used in the programming process for the fixation of a temporary shape. The material is deformed above the thermal transition temperature ($T_{trans}$, determined by differential scanning calorimetry under standard conditions) related to the so-called switching domains between such physical crosslinks and then cooled (while holding the deformation) below $T_{trans}$. Shape recovery is driven by entropy elasticity and occurs on heating the material to a temperature above the switching temperature ($T_{sw}$), which is determined as temperature at maximum recovery rate under the specific conditions of the experiment (e.g. heating rate). In this way $T_{sw}$ differs from $T_{trans}$. Triple-shape polymers consist of at least two types of segregated domains connected via physical or chemical crosslinks with transition temperatures $T_{trans, A}$ and $T_{trans, B}$, so that each domain can be used for the fixation of one temporary shape.

Here, we report structural investigations utilizing SAXS and WAXS on polymer networks based on poly($\varepsilon$-caprolactone) (PCL) and poly(cyclohexyl methacrylate) (PCHMA) segments. These polymer networks exhibit a triple-shape capability, which can be obtained by a one-step programming. For a PCL content between 35 and 60 wt% the polymer networks are phase-segregated and consist of an amorphous PCHMA phase, a crystalline as well as an amorphous PCL phase. This enabled the investigation of morphology-changes related to the PCL crystalline domains during different steps of the SMCP and recovery cycle as well as of the influence of the amorphous components on changes in morphology. Based on parameters determined by SAXS and WAXS, the morphology on the size level of crystalline and amorphous domains and a basic structural model for this multiphase polymer network are discussed.

**EXPERIMENTAL DETAILS**

AB polymer networks based on PCL and PCHMA segments were synthesized from PCL-di-methacrylates (PCLDMA) and cyclohexyl methacrylate (CHMA) by free radical polymerization, induced by irradiation with UV light (Fig. 1). For nomenclature of the networks, the weight

50

fraction of PCLDMA in the co-monomer starting mixture is indicated by a number in parentheses. CHMA (Aldrich) was used without further purification. PCLDMA was obtained by endgroup functionalization of PCL diol (Aldrich) ($M_n \sim 9700$ g $\cdot$ mol$^{-1}$) with methacryloylchloride. Swelling experiments in chloroform were performed to determine the degree of swelling ($Q$), which gives an indirect measure for the crosslink density, i.e. high degree of swelling indicates low crosslinking density. The concentration of the crosslinker was suitable, to give sufficiently crosslinked AB polymer networks for concentrations from PCL(30)CHM to PCL(60)CHM.

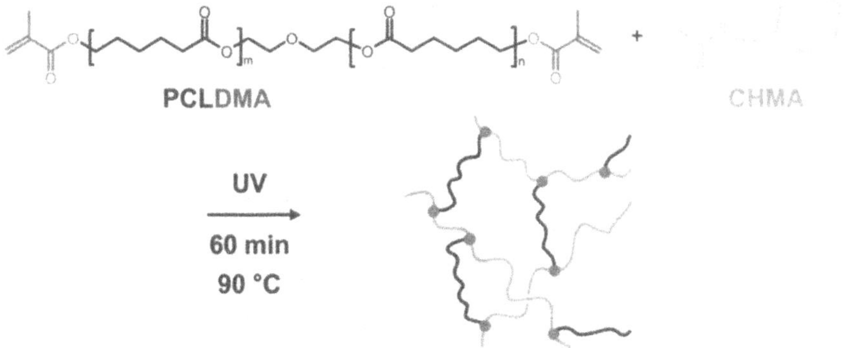

**Figure 1.** Chemical structure of PCLDMA and CHMA and schematic representation of the resulting polymer network structure for PCL/CHM (cp. Ref. 3).

Cyclic, thermomechanical experiments were performed on a Zwick (Ulm, Germany) Z1.0 tensile tester with a 100 N load cell, equipped with a thermochamber W91255 and a Eurotherm 2408 controller [4]. Cooling was realized by a gas flow from a liquid nitrogen tank. Samples were cut to standard dimensions (ISO 527-2/ 1BB). The crosshead speed in all experiments was 10 mm min$^{-1}$. In a stress-controlled cyclic, thermomechanical experiment the sample was strained to an elongation of $\varepsilon_m$ = 50 % at a temperature $T_{high}$ = 150 °C. The sample was kept for 5 min at $\varepsilon_m$ and afterwards cooled under constant stress to $T_{low}$ = -10 °C, with a cooling rate of 5 K min$^{-1}$, upon which the sample length changed to $\varepsilon_l(N)$, where N gives the number of the cycle. After subsequent unloading the length of the sample in the temporary shape $\varepsilon_u(N)$ was obtained. The recovery process was then triggered by heating the sample back to $T_{high}$, upon which the sample length decreased to $\varepsilon_p(N)$. Samples were cooled to room temperature at different steps during such a cyclic, thermomechanical experiment, leading to "structural frozen" samples. For the present study three different samples were chosen, an undeformed, a programmed and a recovered one. These samples were investigated by means of SAXS and WAXS.

WAXS measurements were carried out using the X-ray diffraction system Bruker D8 Discover with a two-dimensional detector from Bruker AXS (Karlsruhe, Germany). The X-ray generator was operated at a voltage of 40 kV and a current of 40 mA, producing Cu K$_\alpha$-radiation with a wavelength $\lambda$ = 0.154 nm. A parallel, monochromatic X-ray beam was provided by a graphite monochromator and a pinhole collimator with an opening of 0.8 mm. WAXS images were collected at a sample-to-detector distance of 15 cm; samples with a thickness of about 1 mm were illuminated for 120 seconds in transmission geometry. SAXS measurements at a wavelength of 0.154 nm were carried out using a laboratory pinhole instrument (NanoStar, Bruker AXS, Karlsruhe, Germany) at a sample-to-detector distance of 105 cm.

## RESULTS and DISCUSSION

For polymer networks with a PCL content between 35 and 60 wt% the recovery curves, determined under stress-free conditions, showed two distinct switching processes, indicating a triple-shape capability. Two different $\overline{T}_{sw}$ (average switching temperature for second to fifth cycle in thermomechanical, cyclic tests) were detected in the recovery curves for PCL(40)CHM at $T_{sw,1} = 54°$ C attributed to the crystalline PCL domains and at $T_{sw,2} = 124$ °C which can be correlated to the amorphous PCHMA domains. The recovery curves were reported elsewhere, as well as the determination of the overall shape fixity and recovery rate, which are higher than 98% [4].

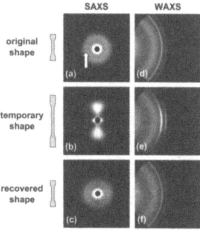

**Figure 2.** Scattering images of a PCL(40)CHM sample at different steps of a programming and recovery cycle, measured at room temperature: (a-c) SAXS images, (d-f) WAXS images.

The SAXS data derived from PCL(40)CHM samples displayed a peak at $Q = 0.38$ nm$^{-1}$, visible as isotropic ring in Fig. 2a (highlighted by a white arrow). This indicated a long period of 16 nm between the crystalline domains, as calculated by the Bragg equation. As a consequence of the programming and resulting elongation, this peak shifted to a value of $Q = 0.44$ nm$^{-1}$ (Fig. 2b), which corresponds to a long period of 14 nm. The peak shifted back to the original position after shape recovery (Fig. 2c). Integration of the data of several samples (SAXS profiles shown in ref. [4]) indicated that the shift of the peak position is statistically significant. In WAXS experiments a diffraction image similar to homo-networks from poly(ε-caprolactone)dimethacrylate was detected [8], showing two pronounced diffraction rings of the orthorhombic PCL crystals, (110) and (200) (Fig. 2d). From the reflection widths of these two strong equatorial reflections an average crystal size of 10 nm in lateral crystallite-direction was determined by use of the Scherrer equation (1).

$$(1) \qquad L_{hkl} = \frac{K \cdot \lambda}{\beta \cdot \cos \theta}$$

In this equation, $L_{hkl}$ is the crystallite size perpendicular to the *hkl* plane, $K$ is the Scherrer constant depending on the crystallite shape, $\lambda$ is the wavelength of radiation, and $\beta$ the integral width of the peak (in radians 2$\theta$) located at an angle $\theta$.

A long period between crystalline domains of 16 nm was determined from the SAXS measurements. The amorphous PCL-domains between the crystalline domains were coarsely estimated to a size of about 6 nm, while the dimension of the amorphous PCHMA-domains could not be determined. The thermomechanical treatment did not influence the PCL WAXS peak widths and positions. As the widths of the equatorial peaks (110) and (200) and therefore the crystal size remained unaffected of the programming process, the crystallization of the stretched samples during the cooling process led to a spatial rearrangement of the crystalline domains with a fiber texture as show in the WAXS image of the programmed sample (Fig. 2e). This was accompanied with a long period of 14 nm for the crystalline domains along the elongation direction and no long period across the elongation direction.

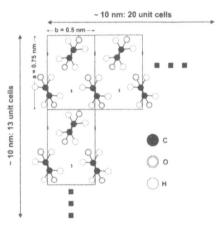

**Figure 3.** Schematic of a PCL crystal in PCL/CHM polymer networks in *ab* projection [cp. Ref. 8]. The upper left corner shows a unit cell and chain arrangement of PCL in the *ab* projection. Around 20 unit cells in *b*-direction lead to a lateral crystal size of 10 nm, while in *a*-direction 13 unit cells are required.

The overall degree of crystallinity (DOC), which here is defined as the ratio between the areas of crystalline peaks and the total area below the diffraction curve (area of crystalline peaks plus area of the amorphous halo), was determined for PCL(40)CHM samples to be around 30 % at room temperature. Assuming –for approximation– a similar scattering efficiency of the two phases, and normalizing to the weight fraction, this results in a DOC of 75 % for the PCL phase.

As the lateral crystallite size was calculated to be roughly around 10 nm, the number of CL repeating units, which form one single crystallite, could be estimated. The lateral dimensions of a PCL unit cell are 0.75 nm in the $a$-direction and 0.5 nm in the $b$-direction (Fig. 3) [8]. In PCL(40)CHM samples the longitudinal dimension ($c$-direction) of the crystallites could not be detected due to a lack of observable meridional peaks. From the number of unit cells in the lateral dimension of one single crystallite (around $20 \cdot 13 = 260$, cp. Fig. 3), with each unit cell containing two CL repeating units, it can be assumed that one crystallite is made up of about 500 PCL chain segments. A maximum approximation of the overall involved repeat units can be obtained by the consideration of the length of a PCL chain segment ($M_n \sim 9700$ g $\cdot$ mol$^{-1}$), which is made up of about 85 CL repeat units ($M_n \sim 114$ g $\cdot$ mol$^{-1}$). A DOC of 75 % for the PCL phase leads to more than 30 000 involved CL repeat units in one average crystal.

## CONCLUSIONS

The investigated covalently crosslinked polymer networks are phase-segregated and consist of an amorphous PCHMA phase and a semi-crystalline PCL phase. In the AB polymer networks different switching phases can be used for the fixation of a temporary shape by an adjustment of the temperature range, in which the experiment is performed. Performing cyclic, thermomechanical tests with $T_{low} = -10$ °C and $T_{high} = 150$ °C enabled both phases to act as switches. Both contributed to the fixation of the temporary shapes, which resulted in a triple-shape effect. The triple-shape capability could be obtained utilizing a one-step dual-shape programming process. Deforming the sample above $T_{high}$, both types of chain segments were oriented and physical crosslinks were formed successively by cooling to $T_{low}$. Cooling led at first to a vitrification of PCHMA followed by a crystallization of PCL. Heating afterwards to 150 °C caused the strain slowly to increase because of thermal expansion of the sample. At about 50 °C the strain slightly decreased which could be explained by melting of PCL crystallites contributing in a certain extent to the fixation of the temporary shape. Further heating led to a recovery of the permanent shape induced by a softening of the glassy PCHMA phase.

While longer PCL segments (with several hundred or more CL repeat units) could fold into lamellar crystals [8], the short PCL segments in the investigated polymer networks with only 85 CL repeat units are assumed to organize in extended-chain crystals. However, the packing of hundreds of chain segments, which are all attached to the network at both of their ends, will lead to large local stresses. Therefore, certainly a considerably lower number than 85 CL repeat units will be involved from each PCL chain segment, while the others will be more or less in the amorphous state. Further studies are required to gain a more accurate description.

## REFERENCES

[1] A. Lendlein, S. Kelch, Angew. Chem. Int. Ed., 41, 2034-2057, 2002.
[2] M. Behl, A. Lendlein, Soft Matter, 3, 1, 58-67, 2007.
[3] I. Bellin, S. Kelch, R. Langer, A. Lendlein, Proc. Natl. Acad. Sci. USA, 103, 18043-18047, 2006.
[4] M. Behl, I. Bellin, S. Kelch, W. Wagermaier, A. Lendlein, Adv. Funct. Mater., 102-108, 2009.
[5] I. Bellin, S. Kelch, A. Lendlein, J. Mat. Chem., 17(28): p. 2885-2891, 2007.
[6] H.S. Gupta, W. Wagermaier et al., Nano Letters 5 (10):2108-2111, 2005
[7] H.S. Gupta, J. Seto, W. Wagermaier et al., Proc. Natl. Acad. Sci. USA, 103, 17741-17746, 2006
[8] H. Bittiger, R. H. Marchessault, W. D. Niegisch, Acta Crystallogr., Sect. B: Struct. Sci., 26, 1923-1927, 1970.

Mater. Res. Soc. Symp. Proc. Vol. 1190 © 2009 Materials Research Society     1190-NN03-21

# Triple-Shape Capability of Thermo-Sensitive Nanocomposites From Multiphase Polymer Networks and Magnetic Nanoparticles

U. Narendra Kumar, K. Kratz, M. Behl and A. Lendlein
Center for Biomaterial Development, Institute for Polymer Research, GKSS Forschungszentrum Geesthacht GmbH, and Berlin-Brandenburg Center for Regenerative Therapies (BCRT), Kantstr. 55, 14513 Teltow, Germany

## ABSTRACT

Thermo-sensitive multiphase polymer networks with triple-shape capability have been recently introduced as a new class of active polymers that can change on demand from a first shape A to a second shape B and from there to a permanent shape C. Such multiphase polymer networks consist of covalent cross-links that determine shape C and at least two phase-segregated domains with distinct thermal transitions $T_{trans,A}$ and $T_{trans,B}$ , that are associated to shape A and B. In general the application of a two step programming or a one step programming procedure is required for creation of triple-shape functionality. In this study we report about a series of CLEGC nanocomposites consisting of silica coated nanoparticles (SNP) incorporated in a multiphase graft polymer network matrix from crystallisable poly(ε-caprolactone) diisocyanatoethyl methacrylate (PCLDIMA) and poly(ethylene glycol) monomethyl ether monomethacrylate (PEGMA) forming crystallisable side chains. These CLEGC nanocomposites were designed to enabling non contact activation of triple-shape effect in alternating magnetic field. Composites with variable PCLDIMA content ranging from 30 wt-% and 70 wt-% and different SNP amounts (0 wt-%, 2.5 wt-%, 5 wt-% and 10 wt-%) were realized by thermally induced polymerization. The thermal and mechanical properties of the CLEG nanocomposites were explored by means of DSC, DMTA and tensile tests. The triple-shape properties were quantified in cyclic, thermomechanical experiments, which consisted of a two step programming procedure and a recovery module under stress-free conditions for recovery of shapes B and C.
While the thermal properties and the Young's modulus of the investigated polymer networks were found to be independent from the incorporated amount of SNP, the elongation at break ($\varepsilon_B$) decreases with increasing nanoparticle content. All investigated composites exhibit excellent triple-shape properties showing a well separated two step shape recovery process.

## INTRODUCTION

Thermo-sensitive multiphase polymer networks capable of a triple-shape effect have been introduced recently as a promising class of active polymers [1-4]. They can change from a first shape (A) to a second shape (B) and from there to a third shape (C), when stimulated by two subsequent increases in temperature. The structural concept for triple-shape polymers based on multiphase polymer networks that are able to form at least two segregated domains. Although the original shape (C) is defined by covalent netpoints resulting from a crosslinking reaction, shapes A and B are created by a two-step thermomechanical programming process. Shape B is determined by physical crosslinks associated to the highest transition $T_{trans,B}$. In the same way, shape A is related to $T_{trans,A}$. Both thermal transitions $T_{trans,B}$ and $T_{trans,A}$ can either be a $T_m$ or $T_g$. In contrast to shape-memory polymers (SMP) with dual shape properties, which are deformed in a single-step procedure [5-10], for the realization of multiple-shape capability the application of a specific multi-step [1-3] is required. Recently, AB polymer networks were introduced, which allowed the creation of triple-shape capability by single-step thermomechanical programming procedure[4]. A characteristic of materials that exhibit the triple-shape effect (TSE) are two distinct switching temperatures ($T_{sw,A \to B}$ and $T_{sw,B \to C}$), which can be determined from the strain-temperature curve under stress-free recovery conditions. The extent, to which the subsequently memorized shapes (B and C) are recovered, is quantitatively described by the strain recovery rates $R_r(A \to B)$ and $R_r(A \to C)$.

In this paper, we explore novel triple-shape polymer nanocomposites (TSPC) prepared by thermal polymerization technique by incorporating SNP [11] in a prepolymer mixture of poly(ethylene glycol) monomethyl ether monomethacrylate (PEGMA: $M_n = 1000$ g·mol$^{-1}$ , $T_m = 38$ °C) and poly(ε-caprolactone) diisocyanatoethyl methacrylate (PCLDIMA: $M_n = 8300$ g·mol$^{-1}$ , $T_m = 55$ °C) acting as crosslinker using benzyl peroxide as initiator. We named the nanocomposites CLEGC. Crosslinking agent PCLDIMA was synthesized from PCL-diol using 2-isocyanatoethyl methacrylate according to ref.[12]. Silica coated nanoparticles were chosen to realize a homogenous distribution of nanoparticles within the polymer matrix without the formation of μm-sized agglomerates. For this study a series of multiphase CLEGC´s with various PCLDIMA weight fractions ranging from 30 wt-% to 70 wt-% and SNP contents from 0 wt-% to 10 wt-% were synthesized. For realization of the triple-shape capability, a specific two step thermomechanical conditioning for creation of triple-shape functionality, which we named triple-shape creation method (TSCM), was applied.

## MATERIALS AND METHODS

### Nanocomposite preparation

The composite preparation is described exemplarily for CLEG040C05. A mixture of 2.8 g of PCLDIMA and 4.2 g of PEGMA (Polysciences, Warrington, PA, USA) was heated to 75 °C, and then 0.37 g SNP (AdNano MagSilica 50, Degussa, Hanau, Germany) were incorporated in the prepolymer mixture by mechanical stirring. This mixture was kept at 80 °C under vacuum (0 mbar) for 30 min to remove volatile components. The thermal polymerization was started by addition of (0.0106 g) benzyl peroxide (Sigma-Aldrich, Taufkirchen, Germany) as initiator under vigorously stirring conditions. This reaction mixture is placed between two glass plates separated with a 1 mm thick PTFE spacer and kept at 80 °C for 24 hours.

### Differential scanning calorimetry

DSC experiments were conducted on a Netzsch DSC 204 Phoenix (Selb, Germany). All experiments were performed with a constant heating and cooling rate of 10 K·min$^{-1}$ and with a waiting period of 2 min at the maximum and minimum temperature. The network samples were investigated in the temperature range from -100 °C to 100 °C, first heated from room temperature to 100 °C, then cooled down to –100 °C and again heated up to 100 °C. Both melting temperatures and glass transition temperature were determined from second heating run.

### Thermal gravimetric analysis

TGA experiments for determination of the incorporated amount of nanoparticles were performed on Netzsch TGA 204 Phoenix (Selb, Germany). All experiments were conducted with a constant heating rate of 20 K·min$^{-1}$. The network samples were heated from 25 °C to 500 °C under $N_2$ atmosphere and then from 500 °C to 900 °C under $O_2$ atmosphere. Temperature was kept constant for 2 min at 500 °C.

## Dynamic mechanical analysis at varied temperature

DMTA measurements were performed on a Gabo (Ahlden, Germany) Eplexor 25 N using standard test specimen (ISO 527-2/1BB) punched from polymer network films. All experiments were performed in temperature sweep mode with constant heating rate of 2 K·min$^{-1}$. The oscillation frequency was 10 Hz. The network samples were investigated in the temperature interval from -100 °C to 100 °C.

## Tensile tests and cyclic, thermomechanical experiments

Tensile tests and cyclic, thermomechanical experiments were carried out on a tensile tester Z005 (Zwick, Ulm, Germany) equipped with a thermochamber controlled by Eurotherm control units (2216E for the Z005, Eurotherm Regler, Limburg, Germany). Load cells suitable to determine maximum forces of 200, 100, and 20 N were used depending on samples and temperature. Films were cut into standard samples (ISO 527– 2/1BB) and strained at an elongation rate of 5 mm·min$^{-1}$.

In cyclic, thermomechanical experiments, the sample was stretched at $T_{high}$ = 70 °C from $\varepsilon_c$ (shape C) to $\varepsilon_B^0$. Cooling with a cooling rate $\beta_c$ = 5 K·min$^{-1}$ to $T_{low}$ = 0 °C under stress-control results in $\varepsilon_B^{load}$ and heating with a heating rate of 2 K·min$^{-1}$ to $T_{mid}$ = 40 °C under stress-free condition lead to $\varepsilon_B$ (shape B). The sample was further stretched to $\varepsilon_A^0$ cooled to $T_{low}$ = 0°C under stress-control with $\beta_c$= 5 K·min$^{-1}$, whereas the elongation decreased to $\varepsilon_A^{load}$. Shape A, corresponding to $\varepsilon_A$, was obtained by unloading after 10 min. The recovery process of the sample is monitored by reheating with a heating rate of 1 K·min$^{-1}$ from $T_{low}$ = 0 °C to $T_{high}$ = 70 °C under stress-free conditions, where the sample recovers the first shape B at $\varepsilon_B^{rec}$ and finally shape C at $\varepsilon_C^{rec}$. This cycle is conducted four times for each sample.


## RESULTS AND DISCUSSION

### Thermal and Mechanical Properties

Thermal properties of the CLEGC were investigated by differential scanning calorimetry (DSC). All CLEGC samples showed a glass transition ($T_g$) in the DSC thermograms around -65 °C, which is attributed to the amorphous PCL domains. This $T_g$ remained constant independent from the composite composition. The CLEGC exhibited two well-separated melting transitions, whereby the melting transition at lower temperatures is related to the crystalline PEG domains ($T_{m,PEG}$) and the melting transition at higher temperatures is attributed to the PCL crystallites ($T_{m,PCL}$). While the $T_{m,PCL}$ remains constant at 50 °C, $T_{m,PEG}$ was found to decrease with increasing PCLDIMA content from 38 °C to 20 °C. The overall heat of fusion decreased with increasing PCLDIMA content from 92 J·g$^{-1}$ to 68 J·g$^{-1}$ (Table 1).

To quantify the impact of the nanoparticles on the mechanical properties of the polymer network, DMTA measurements were conducted. The dynamic mechanical properties of the networks represent the variation of storage modulus (E') and tan δ with temperature. The observed values for $T_{max,\delta}$ attributed to the amorphous PCL domains decreased slightly from -52 °C to -58 °C with increasing crosslink density for all samples. CLEGC showed a systematically decrease of the storage modulus E' with increasing temperature. At temperatures above tan $\delta_{max,}$ $_{PCL}$ the values for E' decreased gradually to a level, which is strongly dependent on the crosslink density, starting around 25 °C. At 40 °C, which is above $T_{m,PEG}$, the storage modulus of the

57

samples started to decrease sharply until the melting of the PCL crystallites is completed at around 60 °C, while a pronounced shift of $T_{m,PCL}$ to higher temperatures was observed with increasing PCLDIMA content.

**Table 1** Sample composition and thermal properties of the triple-shape composites determined by TGA and DSC

| Sample ID [a] | Nanoparticle content[b] [wt-%] | $T_g$ [c] [°C] | $T_{m,PEG}$[c] [°C] | $T_{m,PCL}$[c] [°C] | $\Delta H_m$[c] [J·g⁻¹] |
|---|---|---|---|---|---|
| CLEG(030)C05 | 4.9 ± 0.5 | -65 | 38 | 49 | 92 |
| CLEG(050)C05 | 5.1 ± 0.4 | -64 | 34 | 51 | 82 |
| CLEG(060)C05 | 5.0 ± 0.8 | -65 | 28 | 50 | 74 |
| CLEG(070)C05 | 5.0 ± 0.0 | -66 | 20 | 51 | 69 |
| CLEG(040)C00 | 00 | -66 | 36 | 50 | 91 |
| CLEG(040)C02 | 1.8 ± 0.3 | -66 | 37 | 51 | 82 |
| CLEG(040)C05 | 4.7 ± 0.4 | -64 | 36 | 50 | 85 |
| CLEG(040)C10 | 9.5 ± 1.0 | -66 | 37 | 51 | 73 |

[a] Sample ID the three-digit number in brackets gives the weight content of PCLDIMA in wt% and the last two digits at the end represent the wt-% of added nanoparticles (2.5, 05, 10). [b] Particle content in wt-% determined by TGA [c]. $T_g$, $T_m(PEG)$, $T_m(PCL)$ are the transition temperatures determined by DSC, $\Delta H_m(J \cdot g^{-1})$ is the overall melting enthalpy.

**Table 2:** Mechanical data of CLEGC networks determined at (a) room temperature by tensile tests and for CLEG040C networks at (b) 40 °C and (c) 70°C.

| Sample ID* | Elongation at break( $\varepsilon_B$) [%] | E-Modulus [MPa] |
|---|---|---|
| CLEG(030)C05 [a] | 41 | 30 |
| CLEG(040)C05 [a] | 101 | 26 |
| CLEG(050)C05 [a] | 151 | 46 |
| CLEG(060)C05 [a] | 195 | 72 |
| CLEG(070)C05 [a] | 222 | 102 |
| CLEG(040)C02 [b] | 64 | 18 |
| CLEG(040)C05 [b] | 105 | 21 |
| CLEG(040)C10 [b] | 94 | 17 |
| CLEG(040)C02 [c] | 78 | 1 |
| CLEG(040)C05 [c] | 83 | 1 |
| CLEG(040)C10 [c] | 131 | 1 |

Mechanical testing of the networks was performed under uniaxial stress. The tensile tests were conducted at room temperature, where PCL and PEG domains formed crystalline as well as an amorphous phase and at 40 °C and 70 °C for CLEG(040)C with different SNP content (Table 2). In such tests, only the CLEGC with 70 wt-% PCLDIMA content, containing the highest PCLDIMA amount, showed a distinct yield point followed by a region of plastic deformation, which is the typical behaviour for semicrystalline polymers [13]. For all other networks no yield point was observed. At room temperature, the values for Young´s modulus $E$ were a function of the network composition. The Young´s modulus was found to be almost constant for network compositions with PCLDIMA contents up to 50% and then increasing with growing PCLDIMA amount, while the observed values for $\varepsilon_B$ increased with increasing crosslink density. In general, an increase in Young's modulus ($E$) combined with an increase in $\varepsilon_B$ with increasing crosslink density was observed. It has to be considered that the incorporation of nanoparticles influenced the mechanical properties of the triple-shape networks.

### Triple-Shape Properties

For characterization of the triple-shape effect, a specific cyclic, thermomechanical experiment was developed (see Materials and Methods). In each cycle, the two additional shapes (B and A) were created by a two-step uniaxial deformation, followed by recovering shape B and finally shape C. A typical $\varepsilon$-$\sigma$-T graph obtained from cyclic, thermomechanical tests is shown in Figure 1.

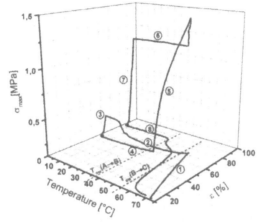

**Figure 1.** Cyclic, thermomechanical experiments of CLEGC(40)C05 (second cycle): (1) deformation at $T_{high}$ = 70 °C to $\varepsilon_B^0$ = 50%, (2) cooling to $T_{low}$ = 0 °C under stress-control, (3) removing the stress at $T_{low}$ = 0 °C, (4) heating to $T_{mid}$ = 40 °C under stress-free conditions, (5) deformation at $T_{mid}$ = 40 °C to $\varepsilon_A^0$ = 100%, (6) cooling to $T_{low}$ = 0 °C under stress-control, (7) removing the stress at $T_{low}$ = 0 °C, (8) heating to $T_{high}$ = 70 °C under stress-free conditions for recovery.

For quantification of triple-shape properties cyclic, thermomechanical experiments were conducted with a series of CLEG040C nanocomposites with different SNP contents. These tests allowed the determination of the shape fixity ratios $R_f(C{\rightarrow}B)$ and $R_f(B{\rightarrow}A)$ as well as the shape recovery ratios $R_r(A{\rightarrow}B)$ and $R_r(B{\rightarrow}C)$ (data are listed in Table 3). $R_f(C{\rightarrow}B)$ is a measure for the fixation of shape B at $T_{mid}$ and $R_f(B{\rightarrow}A)$ for fixation of shape A at $T_{low}$ [14]. Whereas $R_r(A{\rightarrow}B)$ describes to what extent shape B can be recovered starting from shape A [15] and $R_r(A{\rightarrow}C)$ is the overall recovery of shape C starting from shape A. $\overline{R_f}(C \rightarrow B)$, $\overline{R_f}(B \rightarrow A)$,

$\overline{R}_r(A \to B)$ and $\overline{R}_r(A \to C)$ are the average values for cycles 2–4 (the first cycle is used as preconditioning to minimize the contributions of the previous thermomechanical history originating from sample preparation). From the strain vs. temperature curve obtained during stress-free recovery two characteristic switching temperatures $T_{sw,(C \to B)}$ and $T_{sw,(B \to A)}$ could be determined as inflection points.

All investigated composites exhibit excellent triple-shape properties showing an almost complete total recovery $\overline{R}_r(A \to C)$ with two distinct transition steps. $\overline{R}_f(C \to B)$ increased with increasing PCLDIMA content, which supported the fixation of shape B. Accordingly, the contrary tendency was observed for the dependence of $\overline{R}_f(B \to A)$ from the PCLDIMA content. $\overline{R}_r(A \to B)$ values decreased with increasing PCLDIMA content. With the introduction of nanoparticles an influence on the fixity ratio $\overline{R}_f(C \to B)$ and $T_{sw,(B \to C)}$ could be observed.

**Table 3** Triple-shape properties of CLEGC(040)C composites with various SNP content.

| Sample ID | $\overline{R}_f(C \to B)$ [%] | $\overline{R}_f(B \to A)$ [%] | $\overline{R}_r(A \to B)$ [%] | $\overline{R}_r(A \to C)$ [%] | $T_{sw,(A \to B)}$ [°C] | $T_{sw,(B \to C)}$ [°C] |
|---|---|---|---|---|---|---|
| CLEG(040)C00 | 91 | 70 | 52 | 100 | 36 | 52 |
| CLEG(040)C02 | 96 | 70 | 49 | 99 | 37 | 54 |
| CLEG(040)C05 | 94 | 73 | 51 | 99 | 36 | 53 |
| CLEG(040)C10 | 95 | 71 | 58 | 100 | 36 | 55 |

Average values (cycles 2–4) of triple-shape properties determined by cyclic, thermomechanical experiments for $\varepsilon_B^0 = 50\%$ and $\varepsilon_A^0 = 100\%$. The two-digit numbers in parentheses given for the sample IDs are the content of PCLDIMA in the reaction mixture in wt-% and the last two digits at the end are the wt-% of the added nanoparticles.

## CONCLUSIONS

We have demonstrated the realization of triple-shape polymer nanocomposites designed for non contact activation of triple-shape effect in magnetic field via thermally induced polymerization a prepolymer mixture of PCLDIMA and PEGMA and incorporation of silica coated nanoparticles. The particle content was confirmed by TGA measurements. The incorporation of the nano-sized fillers did not influence the thermal properties and the Young's modulus, while a decrease in elongation at break was observed with raising nanoparticle contents. All investigated composites exhibit excellent triple-shape properties with a distinct two step transition. The determined triple-shape characteristics like switching temperatures, fixity ratios and recovery ratios were not influenced by incorporation of the nanoparticles, which enable the activation of such nanocomposites in the magnetic field.

## REFERENCES

1. I. Bellin, S. Kelch, R. Langer and A. Lendlein, Proceedings of the National Academy of Sciences of the United States of America **103** (48), 18043-18047 (2006).
2. I. Bellin, S. Kelch and A. Lendlein, Journal of Materials Chemistry **17** (28), 2885-2891 (2007).
3. I. S. Kolesov and H.-J. Radusch, eXPRESS Polymer Letters **2(7)**, 461-473 (2008).
4. M. Behl, I. Bellin, S. Kelch, W. Wagermaier and A. Lendlein, Advanced Functional Materials **19** (1), 102-108 (2009).

5.  A. Lendlein and S. Kelch, Angewandte Chemie-International Edition **41** (12), 2034-2057 (2002).
6.  M. Behl and A. Lendlein, Soft Matter **3** (1), 58-67 (2007).
7.  M. Behl and A. Lendlein, Materials Today **10** (4), 20-28 (2007).
8.  S Hayashi, Y Tasaka, N Hayashi, Y Akita and Mitsubishi Heavy Industries Technical Review 2004, 1-3.
9.  A. B. Victor, V. N. Varyukhin and V. V. Yurii, Russian Chemical Reviews (3), 265 (2005).
10. M. Yoshida, R. Langer, A. Lendlein and J. Lahann, Polymer Reviews **46** (4), 347-375 (2006).
11. H. Gottfried, Janzen, C., Pridoehl, M., Roth, P., Trageser, B. & Zimmermann, G. (2003) and U.S. Patent 6, 767.
12. S. Lin-Gibson, S. Bencherif, J. A. Cooper, S. J. Wetzel, J. M. Antonucci, B. M. Vogel, F. Horkay and N. R. Washburn, Biomacromolecules **5** (4), 1280-1287 (2004).
13. F. R. Schwarzl, Polymermechanik; Springer: Berlin (1990).
14. Fengkui Li, W. Zhu, X. Zhang, C. Zhao and M. Xu, J. Appl. Polym. Sci. **71** (7), 1063-1070 (1999).
15. H. Tobushi, H. Hara, E. Yamada and S. Hayashi, Smart Materials & Structures **5** (4), 483-491 (1996).

# Stimuli-sensitive Hydrogels

Mater. Res. Soc. Symp. Proc. Vol. 1190 © 2009 Materials Research Society　　　1190-NN03-07

# Mechanically Reinforcing Polyacrylate/Polyacrylamide Hydrogels Through the Addition of Colloidal Particles

Bryan A. Baker[1], Rebecca Murff[2] and Valeria T. Milam[1, 2, 3]
School of Materials Science and Engineering[1], Wallace H. Coulter Department of Biomedical Engineering[2], and Petit Institute for Bioengineering and Bioscience[3], Georgia Institute of Technology, 771 Ferst Drive NW, Atlanta, Georgia 30332-0245

## ABSTRACT

Polyacrylamide is a popular material for many bio-related applications ranging from electrophoretic separation to cellular supports. A limitation of polyacrylamide-based hydrogels, however, is their mechanical compliance. The current study examines the effect of colloidal particles as a reinforcing filler phase to enhance the mechanical stiffness of polyacrylamide-polyacrylate hydrogels. Measurements with oscillatory rheology show that for a fixed polymer volume fraction, the presence of colloidal particles with various surface modifications generally results in an increase of the shear storage modulus of the hydrogel-particle composite. Interestingly, this study indicated that no discernable trends can be linked between the values of the shear storage modulus and the particle surface characteristics.

## INTRODUCTION

Hydrogel networks are comprised of cross-linked hydrophilic polymer chains that form three-dimensional networks[1]. While the high water content provides favorable permeability characteristics, these water-swollen networks are mechanically weaker than many physiological tissues. Past approaches to increase the mechanical stiffness of hydrogels have involved increasing the concentration of cross-linker[2], increasing the monomer concentration[3], altering copolymer composition[4], and embedding particles into the hydrogel[5]. Our work seeks to expand upon these efforts by using colloidal particles as interactive filler additions to polyacrylamide-based hydrogels. Oscillatory rheology is used to determine the effects of particle volume fraction as well as various chemical modifications to the particle surface.

## EXPERIMENT

### Materials

All chemicals were purchased from Sigma (St. Louis, MO) unless otherwise noted. $N,N'$-methylene-bisacrylamide (BIS), $N,N,N',N'$-tetramethylethylenediamine (TEMED), poly(diallyldimethylammonium chloride) (PDDA), and $N$-ethyl-$N'$-(3-dimethylaminopropyl) carbodiimide hydrochloride (EDAC) were used as-received. DIamond filtered nanopure water (Barnstead International) was used to prepare all initiator and buffer solutions. Phosphate buffered saline (PBS) was diluted from a 10x concentration to a 1x concentration. Sodium acrylate and acrylamide were dissolved in PBS to prepare either 40 or 60 wt% monomer solutions. Ammonium persulfate was diluted to a 10 wt% solution and used as the initiator. Spermine solution was prepared by adding 500 mg to 1 mL nanopure water. Carboxylated, 0.79 μm diameter, polystyrene particles (Bangs Laboratories, Fishers, IN) served as the colloidal species.

## Particle Preparation and Zeta Potential

For negatively charged particles, carboxylated particles were used without further modification. For positively charged particles, two methods were used to functionalize carboxylated particles with cationic species: 1) covalent coupling via EDAC of Spermine (MW = 202.34 Da) or 2) adsorption of cationic PDDA (MW < 100,000 Da). For spermine conjugation, 100 μL of particles at 10% solids concentration were centrifuged for 2 minutes at 9,900 g. The supernatant was removed and replaced with 100 μL of PolyLink Coupling Buffer (PLCB) purchased from Bangs Laboratories, Inc (Fishers, IN). The beads were resuspended, centrifuged and again resuspended with 150 μL of PLCB. After the beads were resuspended, 200 μL of PLCB was added to 100 mg of EDAC (pre-weighed and stored in nitrogen backfilled glass vials) then quickly vortexed. 100 μL of the EDAC solution was added to the particle suspension along with 20 μL of spermine solution. After brief mixing, the suspension was mixed end over end for approximately one hour. EDAC and spermine solutions were added two more times followed by overnight mixing.

For PDDA adsorption onto COOH particles, 100 μL of particles were aliquoted and centrifuged as previously described. The supernatant was removed and replaced with 100 μL of PLCB. After a second round of centrifugation and removal of supernatant, the particles were resuspended in 150 μL of PLCB and 20 μL of PDDA was added. The suspension was mixed end over end for approximately one hour. 20 μL of PDDA was added each hour for 2 additional hours followed by overnight mixing.

For zeta potential measurements, 9 v% suspensions of carboxylated, spermine-conjugated, or PDDA-adsorbed particles were prepared in PBS. A 0.5 μL aliquot was then taken from the particle suspensions and added to 1 mL of PBS. In order to quantify the surface charge of the colloid particles, zeta potential measurements were taken on a Malvern Zetasizer Nano Series (UK). The measurements were taken at 37 °C after a 20 minute equilibration time to match rheology conditions.

## Hydrogel-Particle Composite Synthesis and Rheology

Sodium polyacrylate-co-polyacrylamide (pNaAc/pAAm) hydrogels were prepared from monomer solutions of sodium acrylate and acrylamide at a fixed volume ratio and total hydrogel volume fraction, $\Phi_h$, of 0.10. While the ratio of monomers was varied in this study, the concentration of BIS cross-linker was fixed at 0.0024 volume fraction (based on total hydrogel volume and volume of pure BIS). For the polymerization step, 1 v% ammonium persulfate solution and 1 v% TEMED were used. After mixing the monomer solutions with BIS cross-linker, the monomer and initiator solutions were degassed separately via nitrogen bubbling (1 minute and 5 minutes, respectively) to reduce oxygen scavenging of primary radicals formed during initiation[6]. Before adding the initiator and TEMED to the monomer solutions, the prepared polystyrene colloids were dispersed in the monomer solution by vortexing and sonication. Initiator and TEMED were then added and the solution vortexed briefly. While still in liquid form, 0.140 mL of the hydrogel-particle suspension was loaded onto the sample stage of the rheometer. The cone was then lowered to a gap setting of 0.052 mm. The sample was enclosed by a Peltier thermal hood with an evaporation blocker to minimize water loss. The temperature was increased to 37°C for 20 minutes before measurements began.

Oscillatory rheology measurements consisted of strain amplitude sweeps performed on an Anton-Paar MCR 301 rheometer (Anton Paar USA). Strain sweeps were conducted at a frequency of 6 rad/sec and a strain range of 0.001 - 3000%. We report the shear storage modulus, $G'$, at a strain value of 0.025% which is within the linear viscoelastic regime for each hydrogel-particle composite tested. All measurements were done using a cone and plate geometry of 25 mm in diameter (CP-25).

## DISCUSSION

Any potential increase in the mechanical stiffness of polyacrylamide-based hydrogels via the addition of surface-modified colloid particles was postulated to originate from attractive electrostatic interactions between the negative charge groups of the acrylate monomers incorporated into the polyacrylamide hydrogel and the surface charges present on the colloids modified with cationic groups. The average zeta potential values were taken from three separate samples for each functionalized particle and are as follows: -35.77 mV for carboxylated particles, +25 mV for PDDA modified particles, and +23.53 mV for spermine modified particles.

Hydrogel-particle composites were prepared at varying hydrogel compositions (15/85, 50/50, and 100/0 volume percentage ratios of polyacrylate/polyacrylamide) and particle volume fractions, $\Phi_p$, of 0.05, 0.10, and 0.25 and then characterized with oscillatory rheology. For all rheological plots, the data points represent the average of three to six separate samples with shear storage modulus, $G'$, values taken at a strain of approximately 0.025%.

Generally, across all samples tested, an increase in particle volume fraction resulted in a modest increase in $G'$. Figure 1 shows $G'$ as a function of particle volume fraction for hydrogels embedded with carboxylated particles. A pure hydrogel reference (no particles) is included for comparison. Notably, the storage modulus increases for both the 15/85 and 50/50 compositions at the lowest particle volume fraction of 0.05 but does not continually increase at higher particle volume fractions of 0.10 or 0.25. Only a slight increase in $G'$ is observed at these particle volume fractions indicating that the increase in $G'$ is not strongly affected by increasing $\Phi_p$. The 100/0 composition was the most compliant hydrogel tested and expected to show the most dramatic increase in $G'$ with the addition of particles; however, essentially no change in $G'$ is observed.

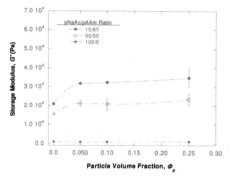

**Figure 1.** $G'$ as a function of particle volume fraction, $\Phi_p$, for hydrogels embedded with carboxylated particles. Hydrogel volume fraction, $\Phi_h$, was fixed at 0.10 with varying pNaAc/pAAm volume percentage ratios.

Figure 2 and Figure 3 show results for hydrogels embedded with oppositely charged particles that were modified with either spermine or PDDA. Figure 2 shows that the addition of spermine-modified particles increased the storage modulus moderately for the 15/85 composition. For the 50/50 composition, the results are difficult to ascertain due to the relatively wide range in measured $G'$ values for the highest particle loading of 0.25. The 100/0 composition is the most compliant hydrogel studied, but again shows little change in $G'$ with particle additions. Overall, a modest increase in $G'$ with the incorporation of spermine-modified particles is apparent.

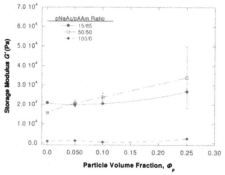

**Figure 2.** $G'$ as a function of particle volume fraction, $\Phi_p$, for hydrogels embedded with spermine-modified particles. Hydrogel volume fraction, $\Phi_h$, was fixed at 0.10 with varying pNaAc/pAAm volume percentage ratios.

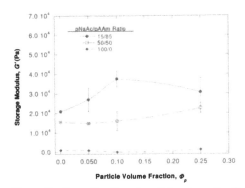

**Figure 3.** $G'$ as a function of particle volume fraction, $\Phi_p$, for hydrogels embedded with PDDA-modified particles. Hydrogel volume fraction, $\Phi_h$, was fixed at 0.10 with varying pNaAc/pAAm volume percentage ratios.

Increasing the particle volume fraction of PDDA-modified particles also resulted in a modest increase in the shear storage modulus for the 50/50 hydrogel composition, but had a more pronounced effect on $G'$ for the 15/85 composition. For the 100/0 composition, little change in $G'$ is observed as particles are added. Since both the spermine and PDDA-modified particles have similar zeta potential values, it is not likely that electrostatic interactions alone result in this difference in mechanical behavior. A more likely influence on the $G'$ behavior of these two systems is due to the size differences of the spermine (~200 Da) and PDDA (< 100,000 Da) macromolecules. The longer PDDA chain segments may more readily entangle with the surrounding hydrogel matrix to provide mechanical reinforcement. To elucidate the possibility of physical entanglement, a microscopy study was conducted (results not shown). Pre-initiated monomer solutions were prepared and a low volume fraction (0.001) of each modified particle was added to see if aggregation occurred in the presence of monomer. Images were taken with a 40x objective on a Zeiss inverted microscope at approximately 15, 30 and 60 minutes after loading. For both the spermine and carboxylate-modified particles there was a predominant population of single particles with some small aggregates of two to five particles present at all compositions after 30 minutes. The PDDA particles, however, showed large clusters consisting of 10's of particles after 30 minutes for the 15/85 monomer composition. Smaller clusters were seen for the 50/50 and 100/0 compositions. Collectively, the results indicate that electrostatic repulsion did not prevent PDDA coated particles from aggregating. Based on these observations it seems reasonable that entanglements of the PDDA-modified particles with the chains of the hydrogel may also occur in the polymerized hydrogel composite.

It is tempting to say that the increase in $G'$ for the 15/85 hydrogel with PDDA-modified particles is simply due to entanglements with the surrounding matrix. However, when comparing the data for all particle embedded 15/85 hydrogels in Figure 4, a different picture emerges. In fact, Figure 4 shows that hydrogels with embedded carboxylated particles have higher $G'$ values at $\Phi_p$ of 0.05 and 0.25. As 15/85 is the least ionized of the copolymer hydrogel composites, it is possible that the weaker net charge of this hydrogel facilitated some of the clustering observed in the particle microscopy study. However, the mechanism behind such clustering cannot be completely resolved between physical entanglement and any ionic effects without further investigation.

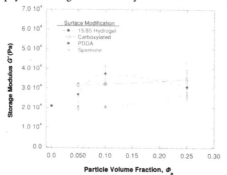

**Figure 4.** $G'$ as a function of particle volume fraction, $\Phi_p$, for hydrogels embedded with carboxylated, spermine-modified, or PDDA-modified particles. Hydrogel volume fraction, $\Phi_h$, was fixed at 0.10 and hydrogel composition was 15/85 pNaAc/pAAm.

## CONCLUSIONS

Increase in $G'$ facilitated by purely electrostatic interactions is not supported by the current study. While entanglements between adsorbed PDDA and the hydrogel network may provide a partial explanation for changes in $G'$, particularly for the 15/85 composition, it does not explain the general increase in $G'$ observed for increasing particle volume fraction for all particles studied. Further research is necessary to resolve these issues and to determine the role of hydrogel charge as a function of composition if any on $G'$ behavior.

## ACKNOWLEDGMENTS

The authors thank Dr. Niren Murthy for helpful discussions and acknowledge funding support from Georgia Tech, ACS (PRF43683-G7). R. Murff acknowledges support from the REU program sponsored by NSF (DMR-0453293).

## REFERENCES

1.    Peppas, N.A., *Hydrogels*. Biomaterials Science: An Introduction to Materials in Medicine: p. 60-64.
2.    Pelham, R.J. and Y.-L. Wang, PNAS, 1997. **94**: p. 13661-13665.
3.    Trompette, J.L., et al., Journal of Polymer Science: Part B: Polymer Physics, 1997. **35**: p. 2535-2541.
4.    Breedveld, V., et al., Macromolecules, 2004. **37**: p. 3943-3953.
5.    Thevenot, C., et al., Soft Matter, 2007. **3**: p. 437-447.
6.    Benda, D., et al., European Polymer Journal, 2001. **37**: p. 1247-1253.

Mater. Res. Soc. Symp. Proc. Vol. 1190 © 2009 Materials Research Society          1190-NN03-09

## Phase Behavior and Shrinking Kinetics of Thermo-Reversible Poly(N-Isopropylacrylamide-2-Hydroxyethyl Methacrylate)

Christine Leon[1], Francisco J. Solis[2] and Brent Vernon[1]

[1]Center for Interventional Biomaterials, Harrington Department of Bioengineering, Arizona State University, Tempe, AZ 85287, USA
[2]Division of Mathematical and Natural Sciences, Arizona State University, Glendale, AZ 85069

**ABSTRACT**

We study the thermodynamic properties of solutions of the physically gelling poly(N-isopropylacrylamide-2-hydroxyethyl methacrylate) [poly(NIPPAm-HEMA)]. We construct its phase diagram and characterize its kinetics of phase separation. This material belongs to a class of thermosensitive, "smart" polymers, that exhibit complex phase behavior.

The copolymer studied is liquid at low temperatures and undergoes phase separation near 28 °C, with negligible dependence on concentration. Above the transition temperature we observe coexistence between a polymer-dilute solution and a gel. We show that, upon quick heating, liquid solutions form a homogeneous gel that phase separates (shrinks) from a dilute polymer solution. We find that the evolution of the gel volume fraction is well described by a double exponential decay, indicating the presence of two shrinking regimes in a close parallel to the behavior of chemically cross-linked gels. The first stage is characterized by quick water ejection. In the second stage, slower shrinking is observed associated with internal reorganization of the polymers that allows the creation of gel-forming contacts.

**INTRODUCTION**

Stimuli-responsive biocompatible polymers have many potential medical applications. An important class of these materials, N-isopropylacrylamide copolymers[1-6], exhibits a Lower Critical Solution Temperature (LCST). Above the LCST, a polymer solution can undergo a transition from a liquid to a two phase coexistence state. This feature allows many of these materials to be injectable at room temperature, while forming gels *in vivo*[1-6]. Possible applications include their use as delivery systems for insulin and cancer drugs, and as implantations to treat aneurysms, arteriovenous malformations, and hypervascularized tumors[1-6].

Since possible key applications of these materials rely on phase transformations, their kinetic and dynamic properties have great importance. While phase separation kinetics in chemically crosslinked gels is well characterized and understood, there currently exists very limited results for thermoreversible gels.

The phase separation process in cross-linked gels has been suitably characterized by means of a double-exponential decay for the volume fraction occupied by the gel[8-10]. The two time scales in the decay reflect two distinct processes: after a sudden temperature change, the gel starts shrinking, ejecting water. The decay rate is set mainly by the friction between the ejected liquid and the gel. Afterwards, a slower rearrangement of the polymers further reduces the volume and improves the favorable local contacts of the polymer. The rate for this second process is determined by the ability of polymer segments to explore new conformation. As we show below, similar behavior is observed in thermoreversible gels. We propose that the microscopic processes at work are similar to those in chemical gels. A scheme of the process appears in Fig. 1. The only major difference between chemical and the thermoversible gels is the ability of the later type to create new crosslinks. Finally, we note that some important properties of the material considered are closely related to its LCST-type behavior. The consequences of this type of phase behavior for gels have been considered in several recent works[11-14].

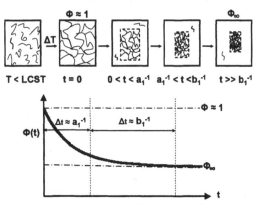

**Figure 1.** Schematic graph of the gel volume fraction evolution $\phi$, and proposed microscopic processes. After heating, an incipient volume filling gel is formed. In the first regime of phase separation, with time scale $a_1^{-1}$, the gel shrinks without creating many more new contacts. In the second regime, with time scale $b_1^{-1}$, smaller shrinkage is observed, but new contacts are formed.


## EXPERIMENT
### Materials
N-isopropylacrylamide (NIPAAm, Spectrum) and was purified by recrystallation in hexanes and vacuum dried for four days. 2,2'-Azobisisobutylronitrile (AIBN, Aldrich) was recrystallized in methanol and filtered. Hydroxyethylmethacrylate (HEMA), diethyl ether, acrylic acid, 1,4-dioxane, tetrahydrofuran (THF), were used as received from Aldrich & Co., St. Louis, MO.
### Polymer Synthesis
Poly(NIPAAm-co-AAc) was synthesized in approximately 10-g batches by free radical polymerization with a 90:10 NIPAAm/HEMA content in 1,4-dioxane with AIBN as the initiator. The monomer content was 10w% in 1,4-dioxane. Oxygen was purged from the solution by bubbling $N_2$, and $7 \times 10^{-3}$ moles of the initiator were then added. The copolymerization was conducted for 18 h at $60^{\circ}C$. The copolymer was precipitated drop-wise by adding excess diethyl ether, collected by vacuum filtration and dried over night under vacuum. The product was then dissolved in deionized water, dialyzed for 3 days against water and lyophilized. The polymer was synthesized by the reaction schematic shown in Figure 2. The synthesized polymer had a mole ratio of NIPAAm to HEMA of 80.6:19.4 and polydispersity index 1.768.

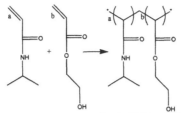

**Figure 2.** Synthesis scheme of Poly(NIPAAM-HEMA).

**Differential scanning calorimetry**

Multicell differential scanning calorimetry (DSC; Calorimetry Sciences Corp.,Lindon, UT)was completed with 5.0 w% solutions of poly(NIPPAm-HEMA) in 0.1M Phosphate Buffered Solution (PBS) with pH values of 7.4. $0.500 \pm 0.005$ g of the solution were placed in DSC ampoules and then tested from 0 to 80°C at 1°C/min in triplicate with PBS as a reference. The Lower Critical Solution Temperatures (LCSTs) were determined by the maximum vertical distance with reference to PBS baseline.

**Turbidity**

1 w% solutions were placed in cells with dimensions of $1 \times 1 \times 5$ cm$^3$ in triplicate using an ultraviolet–visible spectroscopy system (Agilent 8453, UV–vs, HP Co.) with 0.1M, pH 7.4 PBS. The absorbance was measured at a wavelength of 500 nm from 20 to above 40 °C at a scanning rate of 1°C/min with deionized water as the reference.

**Phase Composition at Equilibrium**

Vials were filled with 5w%, 10w%, 15w%, 20w%, 25w%, 30w% and 35w% solutions in triplicate. The solutions were allowed to reach equilibrium, in the water bath for a range of temperatures between 28-55°C. The liquid volume was measure with a syringe and the total volume was found with the height of the total solution and the diameter of the vial.

**Rate to Equilibrium**

Samples of the copolymer in PBS were prepared with concentrations, 10, 15, 25 and 30 w %. Approximately 0.8 mL of each solution were placed vials. The vials were glued to a dish to remain level for imaging. Initially, the dish was at room temperature where the solution was completely sol. Subsequently, the dish was placed in a 40°C water bath. Pictures where take every minute for 10 minutes and then over larger time increments for 7 days.

The pictures were used to infer the gel fraction at different times by comparing the observed gel volume to the total volume of the two phase system. To determine the volume occupied by the gel it was assumed that the horizontal cross sections of the region were circular. We determined the diameters of the cross sections at equally spaced heights of the region by means of pixel counting. Figure 3 presents a set of images of a 10 w% solution as it shrinks over time at 40°C.

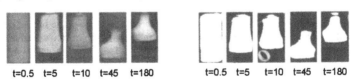

t=0.5  t=5  t=10  t=45  t=180        t=0.5  t=5  t=10  t=45  t=180

**Figure 3.** At left, mages of a 10 w% solution at the indicated times (in minutes). On the right, processed images for inference of cross sections for determination of gel volume fraction.

## RESULTS AND DISCUSSION

Data for the construction of the phase diagram was obtained by allowing phase separated samples to equilibrate for several days. The equilibrium gel volume fraction was determined for samples with initial polymer concentrations of up to 35w% at temperatures from 30°C to 60°C.

The polymer concentration $w_a$ in the phase separated sol-phase was, for all temperatures studied, less than 1w%. To infer the equilibrium concentration in the gel phase, we used data for the gel fraction as a function of weight percentage in the coexistence region. The lever rule indicates that the volume fraction, varies linearly from 0 to 1 in this region. The high concentration boundary, $w_b$, is determined from extrapolation to the $\phi$=1 condition (Figure 4).

The bottom boundary of the phase diagram was inferred through a combination of cloud-measurements and onset of transition in DSC data. The solutions form a gel in the range of temperatures 26-31 °C. The observed range of temperatures is insensitive to polymer concentration. We observed no upper critical solution temperature below 70°C. Results are shown in Figure 4.

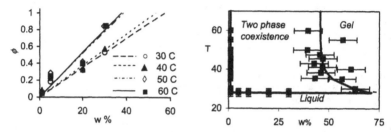

**Figure 4.** (Left) Gel volume fraction versus polymer weight %. Error bars omitted, vertical uncertainties span a range of approximately 10% of the values. (Right) Phase diagram of poly(NIPPAm-HEMA). The left boundary is located at w<1% for all temperatures. Bottom boundary is close to 28°C for all concentrations.

Data for the evolution of the gel fraction of the samples, upon quick heating, was fitted to the double exponential decay form:

$$\varphi(t) = ae^{-a_l t} + be^{-b_l t} + \varphi_\infty \qquad \text{Eq. 1}$$

where $\varphi_\infty = 1\text{-}a\text{-}b$ is the volume fraction of the gel at equilibrium. A sample data set is shown in Fig. 5 (Left). The same data is shown in logarithmic scale in Fig. 5 (Right) along with two separate linear trend-lines for the fast and slow regimes.

Amplitudes $a$ and $b$, and exponents $a_l$ and $b_l$ were fund by fitting the experimental data to the given form by an equally weighted least squares procedure. Plots of these inverse time scales are shown in Figure 5. In general, the results show a larger sensitivity of the second time scale on temperature. This result is consistent with the interpretation of the regime as one associated with thermally activated association processes.

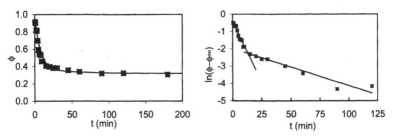

**Figure 5**. (Left) Gel fraction evolution of a 10 w% solution sample heated to 40 °C. The solid line is a fit to the functional form of Eq. 1. (Right) The same data is shown in logarithmic scale with two independent linear fits to the data. Fits to the short and long term behavior clearly exhibits the presence of two different relaxation time scales.

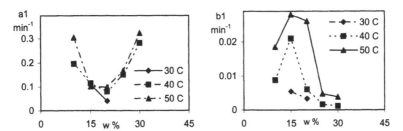

**Figure 6.** Inverse time scale lengths $a_1$ and $b_1$ as function of polymer concentration for different temperatures. Error bars, omitted for clarity, span an average of 10% of the data value.

## CONCLUSIONS

Using the example of the copolymer Poly(NIPAAm-HEMA) we have shown that the kinetics of phase separation in the sol-gel coexistence region of thermoreversible gels can be well described by a double exponential decay, closely resembling the behavior of chemically cross-linked gels. The characterization of the kinetic properties of LCST-type gel forming polymers is a crucial step in the optimization and deployment of these materials in biomedical applications.

## ACKNOWLEDGMENTS

The authors acknowledge funding from the National Institutes of Health, Grant # GM065917 and the Fulton Undergraduate Research Initiative from the Fulton School of Engineering at Arizona State University.

## REFERENCES

1. B. Vernon and A. Martinez, J. Biomat. Sci., Polym. Ed.,16,1153 (2005).
2. B. Jeong and A. Gutowska, Trends. Biotech., 20, 305 (2002).
3. B. Vernon, S.W. Kim, and Y.H. Bae, J. Biomed. Mat. Res., 51, 69 (2000).

4. Z. Cui, B.H. Lee, and B.L. Vernon, Biomacromolecules, 8, 1280 (2007).

5. B. Jeong, Y.H. Bae, D.S. Lee, and S.W. Kim, Nature, 388, 860 (1997).

6. Y. Matsumaru, A. Hyodo, T. Nose, S. Ito, T. Hirano, and S. Ohashi, J. Biomater. Sci., Polym. Ed., 7, 795 (1996).

7. F.S. Bates, Science, 22, 898 (1991).

8. A.N. Semenov, Macromolecules, 31, 1373 (1998).

9. E.E. Dormidontova, Macromolecules, 35, 987 (2002).

10. A. Suzuki, S. Yoshikawa, and G. Bai, J. Chem. Phys. 111, 360 (1999).

11. Y. Maeda, H. Yamauchi, M. Fujisama, S. Sugihara, I. Ikeda, and S. Aoshima, Langmuir, 23, 6561 (2007).

12. F.J. Solis, R. Weiss-Malik, and B. Vernon, Macromolecules, 38, 4456 (2005).

13. M. Watzlawek, C.N. Likos, and H. Löwen, Phys. Rev. Lett., 82, 5289 (1999).

14. M.J. Song, J.H Lee, J.H. Ahn, D.J. Kim, and S.C. Kim, J. Polym. Sci. Part A: Polym. Chem., 42, 772 (2004).

# Biomaterials

Mater. Res. Soc. Symp. Proc. Vol. 1190 © 2009 Materials Research Society          1190-NN06-01

## Design and Realization of Biomedical Devices Based on Shape Memory Polymers

Duncan J. Maitland[1,2], Ward Small IV[2], Pooja Singhal[1], Wonjun Hwang[1], Jennifer N. Rodriguez[1] Fred Clubb[3], and Thomas S. Wilson[2]

[1]Biomedical Engineering Department, Texas A&M University, 3120 TAMU, College Station, TX 77843, U.S.A.
[2]Physical and Life Sciences Directorate, Lawrence Livermore National Laboratory, Livermore, CA 94550, U.S.A.
[3]College of Veterinary Medicine, Texas A&M University, College Station, TX 77843, U.S.A.

## ABSTRACT

Our experience with shape memory polymers (SMP) began with a project to develop an embolic coil release actuator in 1996. This was the first known SMP device to enter human trials. Recent progress with the SMP devices include multiple device applications (stroke treatments, stents, other interventional devices), functional animal studies, synthesis and characterization of new SMP materials, *in vivo* and *in vitro* biocompatibility studies and device-tissue interactions for the laser, resistive, or magnetic-field activated actuators. We describe several of our applied SMP devices.

## INTRODUCTION

Shape memory polymer (SMP) is a material that will have a significant impact on clinical medicine. SMP is a relatively new material that is similar to shape memory alloy (SMA) in its ability to actuate from an initial deformed shape into a second, pre-determined shape. SMP and SMA have material property, fabrication, biocompatibility and cost trade-offs for medical applications. SMP is new to medical applications and has superior shape and volume changing capabilities. The first SMP medical device was just recently approved in March 2009 by the U.S. Food and Drug Administration[1].

SMPs are a class of polymeric materials that can be formed into a specific primary shape, reformed into a stable secondary shape, and then controllably actuated to recover the primary shape. A review of SMP basics and representative polymers was given by Lendlein[2]. Such behavior has been reported in a wide variety of polymers including polyisoprene, styrene-butadiene copolymers[3], segmented polyurethanes[4] and their ionomers[5], copolyesters[3,6], ethylene-vinylacetate copolymers[5], polyacrylamide gels with small amounts of triphenylmethane leucohydroxide[7], and polyacrylic acid[8]. Although there is wide chemical variation in these materials, they can be grouped into categories with high physical similarity based on the method of actuation, which can be achieved thermally, through photo-induced reaction, or by changing the chemical environment[7]. For the SMPs that are actuated thermally, raising the temperature of the polymer above the glass transition temperature ($T_g$) results in a decrease in the elastic modulus from that of the glassy state ($\sim 10^9$ Pa) to that of an elastomer ($\sim 10^6$ to $10^7$ Pa). Upon cooling, the original modulus is nearly completely recovered and the primary form is stabilized. While other classes of thermally activated shape memory materials have been developed, SMAs[9] and shape memory ceramics[10] being the two other major classes[11], SMP has a number of unique and promising properties. Specifically, in the field of medicine, research into the development of new SMPs with adjustable modulus[12], greater strain recovery, an adjustable actuation temperature, and the option of bioresorption[2,13] has opened the possibility for new medical devices and applications.

Some SMP devices that are being researched presently include an occlusive device for embolization in aneurysms[14], a microactuator for removing clots in ischemic stroke patients[15-17], a variety of expandable stents[18] with drug delivery mechanism[19,20], cell seeded prosthetic valves[20], and self-tensioning sutures [2]. One of the key challenges in realizing SMP medical devices is the design and implementation of a safe and effective method of thermally actuating a variety of device geometries *in vivo*. When the soft phase glass transition temperature of the SMP is above body temperature, an external heating mechanism, such as photothermal (laser heating) or electrical (resistive heating) is required[15].

## EXPERIMENT

### SMP Materials

The devices described here were fabricated from two different SMP polyurethane chemistries. The first was thermoplastic block copolymers purchased from DiAPLEX Company, Ltd. This material was used in our earlier experiences, from 1996-2004, with SMP: the microgripper, thrombectomy and stent devices. In 2004 and 2005 we developed our own polyurethane SMPs, nominally labeled LLNL (Lawrence Livermore National Laboratory) materials, and used them in previous device designs as well as in the embolic foam application.

### DiAPLEX SMP

We have previously described processing and the thermomechanical properties of the DiAPLEX SMPs[21,22]. The SMP used in this study was MM7520 (e.g. $T_g$=75 °C), MM5520 and MM3520 polyurethanes obtained from DiAPLEX Company, Ltd. (a subsidiary of Mitsubishi Heavy Industries) as thermoplastic resin in the form of pellets. While the compositions are proprietary, it is known that the material is a segmented polyurethane which has a microphase separated morphology[6]. The primary shape is formed at a temperature above the highest glass ($T_{gh}$) or crystalline ($T_m$) transition and the polymer is cooled in order to fix the shape. Alternatively, the polymer can be dissolved in an appropriate solvent and cast into a mold or coated onto a surface. The secondary shape is obtained by heating the material above the glass transition temperature of the soft phase ($T_{gs}$) or the soft phase crystalline melting temperature ($T_{ms}$), applying a strain to the material, and cooling it down below the same soft phase temperature to fix the shape. The deformation is stored elastically as macromolecular chain orientation within the soft phase, while hard phase segments are relatively unperturbed by this stress. Thus, there is an entropic potential for shape recovery. By heating the material above $T_{gs}$ or $T_{ms}$, soft phase segments spring back to their lower entropic state, resulting in recovery of the primary shape. In vitro biocompatibility of the DiAPLEX materials has been demonstrated[23].

## Custom Polyurethane Chemistry

A description of recipes for several bulk SMP resins was previously described[24]. For devices described subsequently here bulk SMPs were synthesized by adding hexamethylene di-isocyanate (HDI 98% Sigma-Aldrich), N,N,N',N'-tetrakis(2-hydroxypropyl)ethylenediamine (HPED 98% Sigma-Aldrich) and triethanolamine (TEA 99% Sigma Aldrich), in a molar ratio of 10:4:1.3. The ingredients were mixed together vigorously, degassed, and injected into Teflon molds. They were then polymerized 2 hours at 25 °C, ramped to 140 °C at 30 °C/hour, held at 140 °C for 2 hours, then slow cooled. Devices were then stored at room conditions prior to use.

In the case of foams, monomers used included hexamethylene di-isocyanate (HDI 98% Sigma-Aldrich), trimethylhexamethylene di-isocyanate (TMHDI 98% Sigma-Aldrich), isophorone diisocyanate (IPDI, 98% Alfa Aesar), N,N,N',N'-tetrakis(2-hydroxypropyl) ethylenediamine (HPED 98% Sigma-Aldrich) and triethanolamine (TEA 99% Sigma Aldrich). Diisocyanate prepolymers were made to conversions of 35 to 40%. Foams were then made via a combined chemical/physical blowing process using water and Ennovate-3000 (Honeywell International Inc) as blowing agents. Foaming occurred in 1 L polypropylene beakers using a total formulation weight of ~36 g in an oven at 90 °C for 20 minutes under nitrogen, followed by an additional cure for 24 hours at 90 °C under vacuum. Figure 1 shows a SEM image of a 35 °C $T_g$ foam. Foams have been created with $T_g$s ranging from 25 to 90 °C, densities ranging from, approximately, 0.2 g/cc down to 0.015 g/cc, cell sizes ranging from tens of microns up to 3 mm, and cellular structures ranging from mostly closed-cell to nearly fully open-cell.

The unique actuating properties of SMPs can be enhanced further through their structuring into low-density open cell foams [14]. For example, a model equiaxial SMP open-cell foam should have an initial modulus that scales as the square of the solid volume fraction ($\varphi$s) and a yield stress which scales with $\varphi s^{1.5}$ [25]. SMP foam with a solid volume fraction of 0.01 would be expected to have a modulus 0.0001 times that of neat SMP with the promise of similar reduction in recovery stress. At the same time recoverable strains are significantly increased through coupling of structural shape change to molecular orientation relaxations.

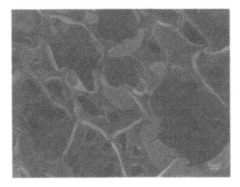

**Figure 1.** SEM image of the Au-Pd sputter coated foam sample as observed at 15kV accelerating voltage, 34 mm working distance and 65X magnification, in a JOEL -JSM 6400 microscope. The image shows formation of numerous membranes indicating the highly porous open-cell structure.

## DISCUSSION

The application of SMP to four different medical device designs is described.

### SMP Microgripper

Our first application of SMP (DiAPLEX thermoplastic) was its use in an actuator designed to hold and, upon thermal activation, release an embolic coil in cerebrovascular aneurysms[25]. This device was developed and tested in clinical trials in 1996. An overview of the catheter system and operational protocol are detailed in Figures 2 and 3. The system consists of a laser, laser driver, display panel, photodetector, fiber coupler, optical fibers, fiber connectors and a SMP-based embolic coil gripper. Figure 4 shows the microgripper releasing an embolic coil in less than 1 second, which was two orders-of-magnitude faster than the release mechanisms used clinically in 1996.

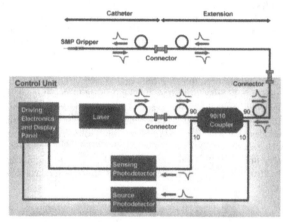

**Figure 2.** A schematic overview of an embolic coil release catheter is shown. Laser light (810nm diode laser, 1.5W) is transmitted through optical fiber, a fiber coupler (90/10 in the schematic), more fiber and into the catheter and gripper release mechanism. This pathway is indicated by the pulses and directional arrows originating at the laser. A small amount of laser light is reflected from the fiber-coil interface back through the coupler into the photodetector. Source fluctuations are monitored by the source photodetector (optional). This pathway is indicated as inverted pulses with directional arrows. As the laser light heats the SMP in the distal tip of the catheter, the gripper releases the embolic coil.

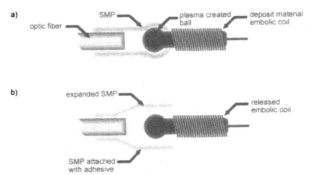

**Figure 3.** Expanded view of the distal tip of the catheter. Figure 3(a) shows the catheter with the coil loaded and ready for deployment. Figure 3(b) shows the catheter after the SMP has been heated above its transition temperature. The SMP around the coil has mechanically relaxed (expanded) to its extruded diameter. The SMP remains in its expanded state after the laser-coupled thermal energy is turned off and the SMP cools below its transition temperature. Note that the SMP around the optical fiber has been permanently attached with an adhesive.

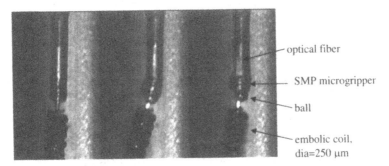

**Figure 4.** The SMP microgripper release actuator is shown releasing an embolic coil (in air). A plasma arc was used to create a ball on the end of the embolic coil. The SMP was mechanically crimped, above $T_g$, around the ball and then cooled. A laser pulse (1 sec, 1.5W) is coupled from an optical fiber into the SMP that has been doped with dye (indocyanine green) to enhance heating by the absorbed laser light. The release of the coil from left to right in the three image series took less than 700 ms.

## SMP Thrombectomy Device

Development of the SMP thrombectomy device was an iterative process based on *in vitro* and *in vivo* functional testing[26-29]. In addition to its ability to capture a blood clot, the fundamental tapered coil design offered several desirable features (see Figures 5 & 6). First, the coil geometry was adaptable for deployment in vessels of various diameters, as it could be fabricated in various sizes; the coil size was simply determined by the mandrel around which the extruded SMP strand was wound while the SMP was programmed. In order to program different coil shapes (number of windings, pitch, diameter of wind and diameter gradient) via different mandrels a thermoplastic (e.g. MM5520) extruded rod (various diameters) was first wound around an aluminum mandrel and held in place with a matching aluminum cap. The final coil shape was programmed into the thermoplastic rod by heating and holding a temperature of 130 °C for 20 minutes. Second, the coil was easily deformed into a straight form for delivery through a catheter prior to thermal actuation. Third, the coil served as a waveguide for laser light, enabling photothermal actuation by coupling an optical fiber directly to the coil. Though a resistive heating mechanism was later adopted, the other features of the tapered coil geometry prompted continued use of the fundamental design.

The original laser-activated SMP coil is shown in Figure 5. It was fabricated completely from commercially available thermoplastic SMP (DiAPLEX)[27,28]. Preliminary benchtop testing in a water-filled silicone vascular model demonstrated feasibility of laser actuation and clot extraction. However, early experiments in rabbits in which blood clots were injected into the carotid artery to simulate an ischemic stroke event revealed two major drawbacks of the original device. First, the length of the straightened coil (~3 cm) inhibited positioning the coil beyond the clot, particularly when the clot was lodged proximal to a bifurcation that required maneuvering the device into a smaller branch vessel. Second, the recovery force of the SMP was not sufficient to achieve complete actuation when the device was pushed up against the vessel wall. This issue was compounded by the coil diameter that was over-sized relative to the narrow rabbit vasculature. Despite the limited efficacy of the original device, several positive results were observed: (1) feasibility of the rabbit acute arterial occlusion model was demonstrated, (2) the device was easily delivered through the guide catheter and visible on fluoroscopy, and (3) the device was able to be retracted back into the guide catheter to re-straighten the coil for repeated extraction attempts and for withdrawal from the body.

Based on the results of the initial animal experiments, the thrombectomy device was re-designed to be slightly smaller and exert more force during shape recovery from the straight to the coil form. The new device was fabricated using a combination of LLNL SMP and SMA (nitinol) as shown in Figure 6[29]. The nitinol provided the necessary shape recovery force required to achieve complete actuation in the narrow rabbit vasculature. The introduction of a nitinol wire backbone within a SMP shell prompted the use of resistive (Joule) heating instead of laser heating for two reasons: (1) the SMP could no longer serve as an efficient waveguide with the embedded nitinol wire in place and (2) the nitinol wire itself could serve as the resistive heating element by passing a current through it. At body temperature, the overlying SMP is in a glassy (high elastic modulus) state and maintains the nitinol coil in a straight form for endovascular delivery. Once in position beyond the clot, Joule heating is initiated. As the surrounding SMP is heated by conduction to its characteristic glass transition temperature ($T_g \approx$ 80 °C), it transitions to its low-modulus rubbery state, allowing the nitinol to resume its coil form. When the current is turned off, the nitinol and the SMP cool and the elastic modulus of the SMP approaches its original glassy value, providing enhanced stiffness to the nitinol coil and resistance to deformation (i.e., stretching) during blood clot extraction.

Following successful retrieval of blood clots in a PDMS vascular model of the rabbit carotid artery, the re-designed SMP-SMA hybrid device was tested in the rabbit acute arterial occlusion model. In the rabbit pilot study, post-treatment angiography showed complete (2/5), partial (2/5), or no (1/5) restoration of blood flow[26]. Figure 7 shows angiographic images acquired during a procedure. Results were similar in the previous benchtop tests. It is notable that the rabbits and model had artery inner diameters of approximately 1-2 mm, which is smaller and more difficult than human cerebral vascular diameters. These results are comparable with the clinical trial investigating the FDA-approved Merci Retrieval System, Concentric Medical Inc. (www.concentric-medical.com), where revascularization occurred in 48% of patients[30]. The Merci device consists of a spring-like nitinol wire that assumes the coil form after being pushed out from a microcatheter positioned distal to the clot (no external heating source is used). Direct benchtop PDMS model comparisons resulted in the SMP-SMA hybrid device outperforming the Merci device. Both devices, however, have a significant shortcoming that, in order to successfully retrieve clots, several centimeters of the device must be pushed distal to the clot.

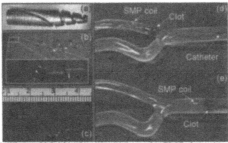

**Figure 5.** As previously shown[27,28], demonstration of a SMP thrombectomy device. (a) Mandrel used to set the primary corkscrew shape of the SMP microactuator. (b) Socket joint between the 200-μm core optical fiber and the SMP coil. The polyimide buffer was burned off and the optical fiber was cleaved prior to insertion into the epoxy-filled socket. The epoxy, whose refractive index was between that of the optical fiber core and that of the SMP, was chosen to provide high coupling efficiency while maintaining a strong bond. (c) SMP coil shown in its secondary straight rod and primary coil forms. The maximum diameter of the SMP coil is approximately 3 mm. (d, e) *In vitro* thrombectomy using the SMP coil device in a bifurcated vessel model. Water flow is from right to left. The SMP microactuator was delivered in its secondary straight rod form through the catheter and pushed distal to the artificial clot (based on acrylamide hydrogel). After laser activation, the SMP microactuator was retracted to capture the clot.

**Figure 6.** As previously shown[29], a hybrid SMP-SMA coil device used in animal clot extraction studies.

**Figure 7.** As previously shown[26], demonstration of *in vivo* clot extraction using the hybrid SMP-SMA coil device. Rabbit carotid angiograms were acquired pre-clot injection, post-clot injection, during device actuation, and post-treatment. Also pictured is the retrieved clot (scale divisions in mm). Though difficult to visualize here, the entire device was clearly visible on the monitors in the operating room (resolution was lost in transferring the images to video tape

## SMP Stents

We developed two prototypes of the stent in addition to solid wall tubes, which were studied for comparison to the machined stents and for animal implantation studies[21,31,32]. The stent fabrication process included dip coating the thermoset SMPs (both DiAPLEX and LLNL SMPs) and laser cutting desired patterns on a cylindrical mandrel. Our initial stent design was optimized for flexibility (Figure 8). This design also turned out to have hoop stresses comparable to commercial alloy stents of the same diameter. The latter design was laser-deployed in a PDMS model artery with and without flow.

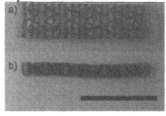

**Figure 8.** As previously shown[32], SMP stent fabricated by laser micromachining. The struts are designed to provide flexibility in both the a) expanded and b) collapsed forms. Scale bar = 1 cm. The outer diameter of the expanded stent is 4.4 mm with a wall thickness of 200 μm.

Compression of the stent pictured in Figure 8 and the solid-walled tubes (not shown) was experimentally studied. The prototype SMP stents used in this study were fabricated from Diaplex MM7520 thermoplastic polyurethane. At 37 °C, both stents exhibited full collapse pressures higher than 4.7 psi, the estimated maximum vasospastic pressure that could collapse an intracranial stent. Full collapse pressure of the laser-machined stent was similar to low collapse pressure stents reported in the literature [33-37]. The stents showed full recovery after crimping, with a radial expansion ratio up to 2.72 and an axial shortening of 1.1 %; higher recovery ratios could be obtained by further dilation.

## SMP Embolic Foams

Recently we have been working on developing SMP embolic foams. The hypothesis is that implanted SMP foams can be used to treat cerbrovascular aneurysms by fully and efficiently filling aneurysms with a clot. The aneurysm is considered non-threatening once the outer clot surface located in the neck of the aneurysm is re-endothelialized, which occurs over the first two weeks post-implantation. The self-expansion, with or without an external energy source, and high volume recovery of SMP foams make them an attractive material for occluding cerbrovascular aneurysms. The SMP foam enables a large volume scaffold on the order of 10s of mm$^3$ to be delivered through the narrow (400 microns, typ) microcatheters in highly tortuous vessels (need for flexibility in the compressed foam).

The repeatability of the foam shape recovery was tested by cycling the compression and expansion of a foam cylinder. LLNL SMP foam block was cut in a cylindrical shape 8 mm in diameter. The cylinder was then cycled through heated compression (Machine Solutions Inc. Crimper), cooling to fix the compressed shape, and expansion using heated air. Figure 9 shows that after an initial 7% loss in expanded diameter, the SMP foam can be repeatedly expanded (35x volume expansion for the foam in Figure 9).

Preliminary aneurysm implantation studies have shown encouraging results. The goal of the foam for embolic procedures is to be mildly thrombogenic where blood contacts the foam but, in the long-term, non-toxic and biocompatible. We are currently working on the long-term studies. Figures 10 and 11 show SEM images of short-term implanted foams in aneurysms. The right carotid artery aneurysm was occluded with foam for one hour and forty-five minutes prior to euthanasia. The left carotid artery aneurysm (not shown) was occluded with foam twenty minutes prior to euthanasia. Harvesting of the aneurysms resulted in isolation of the left and right carotid arteries within 45 minutes of sacrifice. While in place, the aneurysms were fixed with formalin and injected with barium-latex for preservation and visualization with micro computerized tomography respectively. Bisecting the aneurysm and foam down the center, parallel to the center axis of the parent artery, allowed for imaging of the blood and foam interface. This interface was imaged with SEM, shown in Figure 10 below. Figure 10 is an SEM of the foam at 250X magnification, located at the neck of the aneurysm. From these images it can be concluded that the SMP foam promoted thrombogenesis at the site of the aneurysm. Allowing for further time within the animal model may result in full endothelialization of foam and blood interface at the neck of the aneurysm. Figure 11 is a composite of two images taken of the foam and blood interface at the neck of the aneurysm, showing the clotting around the surface of the foam. Upon bisecting the preserved aneurysms and foam, blood had completely penetrated the foam.

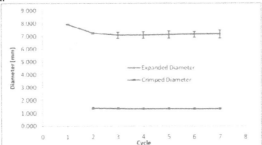

**Figure 9.** Cyclic compression and expansion of SMP foam. Cylindrical SMP foam samples initially 8 mm in diameter were crimped to a diameter of 1.33 mm. This graph shows the average diameter of 3 samples in their expanded and crimped states. Error bars represent the

standard deviation. The standard deviation of the crimped diameter is too small to visualize on this scale.

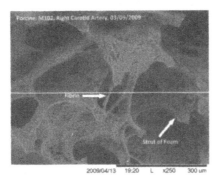

**Figure 10.** Blood and foam interface, 250X magnification. Red blood cells are adhered over the surface of the SMP.

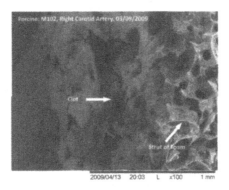

**Figure 11.** Composite image of foam and clot interface, 100X magnification.

## ACKNOWLEDGMENTS

Support provided by the National Institutes of Health/National Institute of Biomedical Imaging and Bioengineering, Grant R01EB000462. We thank Drs. Jonathan Hartman, Matt Miller, Theresa Fossum and Cheng Ji for conducting the aneurysm implantation study. This work was partially performed under the auspices of the U.S. Department of Energy by Lawrence Livermore National Laboratory under Contract DE-AC52-07NA27344.

## REFERENCES

1    M.N. Melkerson, Food and Drug Administration 510(k) approval of Medshape Solution's SMP shoulder anchor, http://www.fda.gov/cdrh/pdf8/K083792.pdf (2009).
2    A. Lendlein and R. Langer, Science **296** (5573), 1673 (2002).
3    T. Takahashi, N. Hayashi, and S. Hayashi, Journal of Applied Polymer Science **60** (7), 1061 (1996).
4    B. K. Kim, S. Y. Lee, and M. Xu, Polymer **37** (26), 5781 (1996).
5    W. Kuhn, A. Katchalsky, and H. Eisenberg, Nature **165**, 514 (1950).
6    S. Hayashi and H. Fujimura, U. S. Patent No. 5049591 (Sept. 17 1991).
7    B. D. Ratner and A. S. Hoffman, in *Introduction to Materials in Medicine*, edited by B. D. Ratner, A. S. Hoffman, F. J. Schoen et al. (Academic Press, New York, 1996).
8    R. P. Kusy and J. Q. Whitley, Thermochimica Acta **243** (2), 253 (1994).
9    L. M. Schetky, Scientific American **241** (5), 74 (1979).
10   M. V. Swain, Nature **322** (6076), 234 (1986).
11   K. Otsuka and C. M. Wayman, *Shape Memory Materials*. (Cambridge University Press, New York, 1998).
12   Yiping Liu, Ken Gall, Martin L. Dunn, Patrick McCluskey, and Robin Shandas, Advanced Materials and Processes **161** (12), 31 (2003).
13   M. Bertmer, A. Buda, I. Blomenkamp-Hofges, S. Kelch, and A. Lendlein, Macromolecules **38** (9), 3793 (2005).
14   A. Metcalfe, A. C. Desfaits, I. Salazkin, L. Yahia, W. M. Sokolowski, and J. Raymond, Biomaterials **24** (3), 491 (2003).
15   D. J. Maitland, M. F. Metzger, D. Schumann, A. Lee, and T. S. Wilson, Lasers In Surgery And Medicine **30** (1), 1 (2002).
16   W. Small, IV, T.S. Wilson, W.J. Benett, J.M. Loge and D.J. Maitland, Optics Express **13** (20), 8204 (2005).
17   M. F. Metzger, T. S. Wilson, D. Schumann, D. L. Matthews, and D. J. Maitland, Biomedical Microdevices **4** (2), 89 (2002).
18   K. Gall, C. M. Yakacki, Y. Liu, R. Shandas, N. Willett, and K. S. Anseth, J Biomed Mater Res A **73** (3), 339 (2005).
19   H. M. Wache, D. J. Tartakowska, A. Hentrich, and M. H. Wagner, J Mater Sci Mater Med **14** (2), 109 (2003).
20   Ken Gall, Christopher M. Yakacki, Yiping Liu, Robin Shandas, Nick Willett, and Kristi S. Anseth, Journal of Biomedical Materials Research - Part A **73** (3), 339 (2005).

21   G. Baer, T. S. Wilson, D. L. Matthews, and D. J. Maitland, Journal of Applied Polymer Science **103** (6) (2007).

22   G. M. Baer, T. S. Wilson, W. Small IV, J. Hartman, W. J. Benett, D. L. Matthews, and D. J. Maitland, Journal of Biomedical Materials Research Part B: Applied Biomaterials (2008).

23   M. Cabanlit, D. Maitland, T. Wilson, S. Simon, T. Wun, M. E. Gershwin, and J. Van de Water, Macromolecular Bioscience **7** (1) (2007).

24   T. S. Wilson, J. P. Bearinger, J. L. Herberg, J. E. Marion III, W. J. Wright, C. L. Evans, and D. J. Maitland, Journal of Applied Polymer Science **106** (1) (2007).

25   D. J. Maitland, A. Lee, D. Schumann, and L. B. Da Silva, U. S. Patent No. 6,102,917 (Aug 15 2000).

26   J. Hartman, W. Small IV, T. S. Wilson, J. Brock, P. R. Buckley, W. J. Benett, J. M. Loge, and D. J. Maitland, American Journal of Neuroradiology **28** (5), 872 (2007).

27   W. Small, M. F. Metzger, T. S. Wilson, and D. J. Maitland, IEEE Journal of Selected Topics in Quantum Electronics **11** (4), 892 (2005).

28   W. Small, T. S. Wilson, W. J. Benett, J. M. Loge, and D. J. Maitland, Optics Express **13** (20), 8204 (2005).

29   W. Small, T. S. Wilson, P. R. Buckley, W. J. Benett, J. M. Loge, J. Hartman, and D. J. Maitland, IEEE Trans Biomed Eng (2006).

30   W. S. Smith, G. Sung, S. Starkman, J. L. Saver, C. S. Kidwell, Y. P. Gobin, H. L. Lutsep, G. M. Nesbit, T. Grobelny, M. M. Rymer, I. E. Silverman, R. T. Higashida, R. F. Budzik, and M. P. Marks, Stroke **36** (7), 1432 (2005).

31   D. J. Maitland, W. Small, J. M. Ortega, P. R. Buckley, J. Rodriguez, J. Hartman, and T. S. Wilson, Optics Express (2007).

32   G. M. Baer, W. Small, T. S. Wilson, W. J. Benett, D. L. Matthews, J. Hartman, and D. J. Maitland, BioMedical Engineering OnLine **6** (1), 43 (2007).

33   W. Schmidt, P. Behrens, D. Behrend, and K-P Schmitz, Prog Biomed Res **4** (1), 52 (1999).

34   K-P Schmitz, W. Schmidt, P. Behrens, and D. Behrend, Prog Biomed Res **5** (3), 197 (2000).

35   R. Rieu, P. Barragan, C. Masson, J. Fuseri, V. Garitey, M. Silvestri, P. Roquebert, and J. Sainsous, Catheter Cardiovasc Interv **46** (3), 380 (1999).

36   C. M. Agrawal, K. F. Haas, D. A. Leopold, and H. G. Clark, Biomaterials **13** (3), 176 (1992).

37   S. Venkatraman, T. L. Poh, T. Vinalia, K. H. Mak, and F. Boey, Biomaterials **24** (12), 2105 (2003).

Mater. Res. Soc. Symp. Proc. Vol. 1190 © 2009 Materials Research Society        1190-NN03-22

# Shape-Memory Properties of Radiopaque Micro-Composites From Amorphous Polyether Urethanes Designed for Medical Application

J. Cui, K. Kratz and A. Lendlein
Center for Biomaterial Development, Institute of Polymer Research, GKSS Research Center Geesthacht GmbH and Berlin-Brandenburg-Center for Regenerative Therapies (BCRT), Kantstrasse 55, 14513 Teltow, Germany

## ABSTRACT

Biocompatible shape-memory polymers are of high significance for application in medical devices or instruments for minimally invasive surgery. To follow the medical device placement or changes in shape of the device in vivo by imaging methods like X-ray techniques, radiopacity of the polymer is required. In this work, we explored the shape-memory properties of radiopaque polymer composites prepared by incorporation of barium sulphate micro-particles in a biomedical grade polyether urethane (PEU) by co-extrusion technique. The filler content was varied from 5 wt% to 40 wt%, which was confirmed by thermal gravimetric analysis (TGA) measurements, while the particle distribution was visualized by scanning electron microscopy (SEM). The thermal and mechanical properties of the composites were investigated by means of dynamic mechanical analysis at varied temperature (DMTA) and tensile tests. The shape-memory properties of PEU composites were quantified in cyclic, thermomechanical experiments.

A significant increase in Young's modulus and a decrease in elongation at break were observed for PEU composites with increasing content of $BaSO_4$, while the DMTA results were not affected by incorporation of the fillers. All samples exhibited excellent shape-memory properties with shape fixity rates ($R_f$) above 98% and values for shape recovery rate ($R_r$) in the range of 81% to 93%. The maximum stress ($\sigma_{max}$) obtained under constant strain recovery conditions increased from 0.6 MPa to 1.4 MPa with raising amount of $BaSO_4$, while the corresponding temperature ($T_{\sigma,max}$) as well as the switching temperature ($T_{sw}$) determined under stress-free conditions remained constant for all polymer composites.

## INTRODUCTION

Shape-memory polymers (SMP) have attracted tremendous interest due to their substantial innovation potential in different application areas, e.g. as intelligent implant materials in medicine [1-3] or in smart textiles [4]. An important class of thermoplastic shape-memory polymers are polyurethanes, which have been applied in artificial heart, heart valves, pacemaker connectors, catheters, vascular stents, suture materials and matrices for controlled drug release [5]. For numerous medical applications inside the human body it is necessary to control the device placement by imaging techniques like X-ray. The most common technique to realize X-ray contrast properties for polymers is the incorporation of radiopaque fillers like barium sulphate ($BaSO_4$).

We explored the fabrication of radiopaque polymer composites from a biocompatible polyether urethane (PEU) with shape-memory capability and $BaSO_4$ micro-particles with various filler contents. The thermal and mechanical properties of PEU composites were analyzed by DMTA and tensile tests. Shape-memory properties of the polymer composites were quantified in cyclic, thermomechanical experiments, using a heating-cooling-heating protocol, while two different recovery modules for recovery of the original shape under stress-free conditions as well as under constant strain conditions were applied.

## EXPERIMENTAL DETAILS

Aliphatic polyether urethanes (PEU) were purchased from Neveon Thermedics Polymer Products (Wilmington, MA, USA) and were used without further purification. The amorphous multiblock copolymer was synthesized from methylene bis($p$-cyclohexyl isocyanate) ($H_{12}$MDI), poly(tetramethylene glycol) (PTMEG, $M_n$ = 1000 g·mol$^{-1}$) and 1,4-butanediol (1,4-BD), which consisted of $H_{12}$MDI / 1,4-BD hard domains with a $T_{g,hd}$ at around 130 °C to 150 °C as well as $H_{12}$MDI / PTMEG soft domains whose $T_{g,sd}$ was reported in the range of -47 °C [6]. Both segments also formed a mixed phase with a pronounced $T_{g,mix}$ which was supported by the urethane-units. The mixed phase mainly determined the thermal and mechanical (*e.g.* Young's modulus) properties of the copolymer and was reported to be suitable as switching segment for shape-memory functionalization [7]. BaSO$_4$ micro particles were purchased from Fagron (Barsbuettel, Germany) and used as received.

Granules of PEU composites with 5, 20, 40 wt% of BaSO$_4$ were prepared via a twin screw extrusion using a EuroPrismLab extruder (Thermo Fisher Scientific). Before processing, the materials were dried in a vacuum oven at 130 °C for 30 min. The processing temperature was in the range between 165 °C and 185 °C, and the extrusion was performed at screw speeds of 25, 50 and 150 rpm. HAAKE Minijet-Micro injection moulding machine (Thermo Electron Corporation) was applied to prepare specimens with type DIN EN ISO 1BB ($I_0$ = 20 mm, width = 2 mm, thickness = 1 mm), which were used for dynamic mechanical analysis at varied temperature (DMTA), tensile tests, and cyclic, thermomechanical measurements. The mould temperature was maintained at 25 °C and the cylinder temperature was 210 °C for PEU and 225 °C for PEU composites.

Tensile tests were carried out on a Z005 tensile tester (Zwick, Ulm, Germany) at 25 °C and the strain rate was 5 mm·min$^{-1}$. For each sample 5 measurements were conducted.

DMTA measurements were performed on an Eplexor® 25 N (Gabo, Ahlden, Germany) equipped with a 25 N load cell. The static load was 10 N, the dynamic load was 3 N, and the oscillation frequency was 10 Hz. All of the experiments were performed in temperature sweep mode from -100 °C to 200 °C with a constant heating rate of 2 K·min$^{-1}$. $T_{\delta,max}$ was determined at the peak maximum of $tan\delta$ - temperature curve.

The weight content of micro-particles was determined by thermal gravimetric analysis (TGA) with temperature ranging from 25 °C to 600 °C. For each sample, the TGA measurement was repeated three times.

The distribution of BaSO$_4$ particles in polymer matrix was analyzed by SEM on a Gemini Supra 40 VP (Zeiss, Oberkochem, Germany) equipped with a RBSD detector at a voltage of 4 kV under low vacuum.

Cyclic, thermomechanical tests were performed on a tensile tester Zwick Z1.0 (Zwick, Ulm, Germany) equipped with a thermo chamber and temperature controller (Eurotherm Regler, Limburg, Germany). Each cycle consisted of a programming module, where the temporary shape was created and a recovery module, where the shape-memory properties under stress-free conditions or under constant strain conditions were determined. Every recovery module was completed by a waiting period of 10 minutes. Three cycles were carried out for each test. In all thermomechanical tests the first cycle was applied as pre-conditioning procedure, while the data was determined in the $2^{nd}$ and $3^{rd}$ cycles. The detailed information about the programming steps is described below:

At the beginning of the $2^{nd}$ cycle the specimen were cooled from $T_{high}$ = 80 °C to $T_{deform}$ = 60 °C. The sample was deformed to $\varepsilon_m$ = 150% at $T_{deform}$ and the strain was kept constant for 5 min to allow relaxation. Afterwards, the sample was cooled down to $T_{low}$ = 0 °C and equilibrated for 10 min at $T_{low}$. Finally, the stress was removed. The elongation of the sample in the temporary state was determined as $\varepsilon_u$.

Recovery under stress-free conditions (recovery module A): the recovery of the original shape was induced by heating the programmed sample to $T_{high}$ = 80 °C with a heating rate of 1 K·min$^{-1}$. The tension-free strain value after the completion of the recovery process was $\varepsilon_p$. Finally, the sample was cooled down to $T_{deform}$, before the next cycle started.

Recovery under constant strain conditions (recovery module B): the sample was heated up to $T_{high}$ = 80 °C for the samples programmed at $T_{deform}$ with a heating rate of 1 K·min$^{-1}$. The strain level was kept constant after programming, so that the length of the sample cannot change during the heating process. The recovery stress built up with increasing temperature until the softening of the material allowed stress relaxation. After reaching $T_{high,}$ the strain constraint was removed allowing the sample to recover its original shape and then the sample was cooled down to $T_{deform}$, before the next cycle started.

## RESULTS AND DISCUSSION

A homogeneous distribution of BaSO$_4$ micro-particles in amorphous polymer matrix was observed in SEM investigations (data not shown). The size of the embedded particles was found to be below 3 μm. TGA measurements confirmed the weight content of BaSO$_4$ particles with values of 5.9 wt% for PEU005, 21.8 wt% for PEU020, and 46.9 wt% for PEU040 (table I)

### Thermal and mechanical properties

In DSC experiments we observed only one $T_g$ for all investigated samples around 55 °C, which is in agreement with the results reported for similar polymer systems [8]. To distinguish between the thermal behavior of the hard and soft segment domains of different PEU composites we performed DMTA measurements. Three relaxation processes were observed in the *tanδ* - temperature curve. The first relaxation process occurred in the range from -80 °C to -10 °C and is attributed to $T_{g,sd}$ of the soft domains (H$_{12}$MDI / PTMEG segment) [9]. The second relaxation process with a pronounced peak maximum $T_{g,mix}$ reflects the glass transition of the mixed phase of H$_{12}$MDI / BD and H$_{12}$MDI / PTMEG segments. At high temperatures between 130 °C and 150 °C, a third transition, $T_{g,hd}$, was found, which is related to the softening temperature of the hard domains (H$_{12}$MDI / BD segments) [6]. As shown in table I, the thermal transitions of the polymer are not influenced by the incorporation of particles. Mechanical properties were investigated by conducting tensile tests at 25 °C. A yield point in the stress-strain curve was observed for all tested samples. The values of Young's modulus increase as the concentration of BaSO$_4$ increases, while the elongation at break decreases (table I). The observed reinforcement by loading PEU with BaSO$_4$ particles can be explained by apparent nucleating effects of BaSO$_4$ for PEU which was also detected for other polymer composites filled with BaSO$_4$ [10].

**Table I.** Composition, thermal and mechanical properties of PEU composites determined by TGA, DMTA and tensile tests

| Sample ID [a] | BaSO$_4$ [b] [wt%] | $T_{g,mix}$ [c] [°C] | $E$ [d] [MPa] | $\sigma_b$ [e] [MPa] | $\varepsilon_b$ [f] [%] |
|---|---|---|---|---|---|
| PEU000 | 0 | 67 | 115 ± 23 | 38 ± 4 | 370 ± 50 |
| PEU005 | 5.9 ± 0.2 | 66 | 248 ± 19 | 40 ± 3 | 360 ± 20 |
| PEU020 | 21.8 ± 1.9 | 73 | 310 ± 8 | 44 ± 2 | 340 ± 30 |
| PEU040 | 46.9 ± 2.0 | 70 | 377 ± 36 | 31 ± 2 | 210 ± 20 |

a). The three-digit number gives the content of particles in wt%; b). The weight content of BaSO$_4$ was determined from TGA. The denoted errors are standard deviations from three measurements; c). $T_{g,mix}$ is the peak maximum from the second relaxation process in the *tanδ* - temperature curve of DMTA measurement; d). $E$ is the Young's modulus detected from tensile tests at 25 °C; e). $\sigma_b$ is the stress at break; f). $\varepsilon_b$ is the elongation at break.

### Shape-memory properties

Shape-memory properties of PEU and its composites were investigated according to the programming procedures mentioned in the experimental details and shown in Figure 1.

Suitable values for quantification of shape-memory capability are shape fixity rate ($R_f$) and shape recovery rate ($R_r$). The shape fixity rate describes how well the material can fix the maximum deformation $\varepsilon_m$ in the temporary shape and is defined as the ratio of the elongation $\varepsilon_u$ and $\varepsilon_m$. The shape recovery rate represents the ability of the material to recover its original shape and is calculated from the strain value of the temporary shape $\varepsilon_u(N)$ and the strain values of the tension-free states $\varepsilon_p(N-1)$ and $\varepsilon_p(N)$ while expanding the sample in two subsequent cycles $N-1$ and $N$ (Equation 1) [11].

$$R_r(N) = \frac{\varepsilon_\mu(N) - \varepsilon_p(N)}{\varepsilon_\mu(N) - \varepsilon_p(N-1)} \qquad \text{Eq. (1)}$$

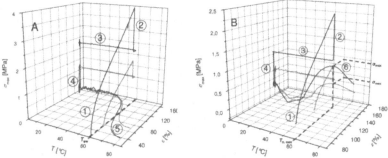

**Figure 1.** Cyclic, thermomechanical experiment of PEU000 (grey) and PEU020 (black); results from the 2$^{nd}$ cycle are shown: A. recovery under stress-free conditions; B. recovery under constant-strain conditions. (1) deformation to $\varepsilon_m = 150\%$ at $T_{deform} = 60$ °C; (2) relaxation for 10 min at 60 °C; (3) cooling down to $T_{low} = 0$ °C; (4) removing the stress at 0 °C; (5) recovery under stress-free conditions or (6) constant strain conditions with $T_{high} = 80$ °C.

In addition to $R_f$ and $R_r$ the characteristic switching temperature $T_{sw}$ can be determined as inflection point from the strain-temperature recovery curves under stress-free conditions, and the maximum stress $\sigma_{max}$ generated during heating as well as its corresponding temperature $T_{\sigma,max}$ are determined from the stress-temperature recovery curve under constant strain conditions.

In all shape-memory experiments, an almost complete fixation of the temporary shape with values for $R_f > 98\%$ was obtained. The values of $R_r$ for PEU composites detected under stress-free conditions were in the range of 81% to 93% in the 2nd cycle and 92% to 97% in the 3rd cycle, and a $T_{sw}$ at 60 °C for all the cycles, indicating the excellent shape-memory properties of PEU composites. In experiments under constant strain conditions an increase in $\sigma_{max}$ was generated from 0.6 MPa to 1.4 MPa in the 2nd cycles for PEU composites with increasing amount of BaSO$_4$. The obtained temperature related to the stress maximum $T_{\sigma,max}$ was independent from the composition of polymer composites around 60 °C (table II).

**Table II.** Shape-memory properties of PEU composites determined in cyclic, thermomechanical experiments

| Sample ID [a] | No. of cycle | $R_f$ [b] [%] | $R_r$ [c] [%] | $T_{sw}$ [d] [°C] | $R_f$ [b] [%] | $R_r$ [c] [%] | $T_{\sigma,max}$ [e] [°C] | $\sigma_{max}$ [e] [MPa] |
|---|---|---|---|---|---|---|---|---|
| | | | Recovery module A | | | Recovery module B | | |
| PEU000 | 2 | 100 | 89 | 60 | 100 | 88 | 61 | 0.4 |
| | 3 | 100 | 97 | 60 | 100 | 94 | 61 | 0.4 |
| PEU005 | 2 | 100 | 81 | 60 | 100 | 81 | 60 | 0.6 |
| | 3 | 100 | 92 | 61 | 100 | 88 | 60 | 0.5 |
| PEU020 | 2 | 100 | 93 | 59 | 98 | 85 | 61 | 0.9 |
| | 3 | 100 | 95 | 60 | 100 | 92 | 60 | 0.9 |
| PEU040 | 2 | 100 | 93 | 60 | 100 | 85 | 60 | 1.4 |
| | 3 | 100 | 96 | 59 | 100 | 93 | 61 | 1.2 |

a). The three-digit number gives the content of particles in wt%; The cyclic, thermomechanical tests were performed with fixed parameters: $T_{deform} = 60$ °C, $T_{low} = 0$ °C, $T_{high} = 80$ °C and $\varepsilon_m = 150\%$; b). $R_f$: shape fixity rate; c). $R_r$: shape recovery rate; d). $T_{sw}$: switching temperature; e). $\sigma_{max}$: maximum stress and its corresponding temperature ($T_{\sigma, max}$).

## CONCLUSIONS

Radiopaque polymer composites were fabricated by co-extrusion from amorphous multi-block copolymers (PEU) and various amounts of BaSO$_4$ micro-particles as radiopaque fillers. The content of BaSO$_4$ was confirmed by TGA and a homogeneous distribution of BaSO$_4$ particles was visualized by SEM. According to DMTA measurement, no impact of the particle content on the thermal behaviour of PEU composites was observed, while an increase in Young's modulus and a decrease in elongation at break were observed with raising filler contents. The PEU composites exhibited excellent shape-memory properties determined under stress-free conditions as well as under constant strain conditions, which are similar to pure PEU samples. While $T_{sw}$ and $T_{\sigma,max}$ were found to be independent from the composition of the polymer composites, the values of $\sigma_{max}$ increased with increasing concentration of BaSO$_4$ particles.

## ACKNOWLEDGMENTS

The authors are grateful to Deutsche Forschungsgemeinschaft for financial support under grant No. SFB760, Project B3.

## REFERENCES

1. A. Lendlein, R. Langer, *Science* **2296,** 1673 (2002).
2. A. Lendlein, S. Kelch, *Clin. Hemorheol. Microcirc.* **32,** 105 (2005).
3. A. Metcalfe, A. C. Desfaits, I. Salazkin, L. Yahia, W. M. Sokolowski, J. Raymond, *Biomaterials* **24,** 491 (2003).
4. J. Hu, X. Ding, X. Tao, J. Yu, *J. Dong Hua Uni. (Engl. Ed.)* **19,** 3 (2002).
5. N. R. James, J. Philip, A. Jayakrishnan, *Biomaterials* **27,** 160 (2006).
6. R. E. Solís-Correa, R. Vargas-Coronado, M. Aguilar-Vega, J. V. Cauich-Rodríguez, J. San Román, A. Marcos, *J. Biomater. Sci. Polymer Edn* **18,** 561 (2007).
7. R. Mohr, K. Kratz, T. Weigel, M. Luka-Gabor, M. Moneke, A. Lendlein, *Proc. Nat. Acad. Sci. U.S.A.* **103,** 3540 (2006).
8. J. R. Lin, L. W. Chen, *J. Appl. Polym. Sci* **69,** 1575 (1998).
9. J. W. Cho, Y. C. Jung, Y. C. Chung, B. C. Chun, *J. Appl. Polym. Sci* **93,** 2410 (2004).
10. X. Guo, *J. Appl. Polym. Sci* **109,** 4015 (2008).
11. A. Lendlein, S. Kelch, *Angew. Chem. Int. Ed.* **41,** 2034 (2002).

Thermomechanical behaviour of biodegradable shape-memory polymer foams

Samy A. Madbouly[*], Karl Kratz, Frank Klein, Karola Lützow, and Andreas Lendlein

Centre for Biomedical Development, Institute of Polymer Research, GKSS Research Centre Geesthacht GmbH, Kantstr.55, 14513 Teltow, Germany

ABSTRACT

Shape-memory polymer foams based on poly($\omega$-pentadecalactone) (PPDL) and poly($\varepsilon$-caprolactone) (PCL) multiblock copolymer with 60 wt% PCL content were prepared by environmentally-friendly high pressure supercritical carbon dioxide (scCO$_2$) foaming technique. A foam with a density of approximately $0.11 \pm 0.02$ g/cm$^3$ and an average pore size of 150-200 μm with excellent compressibility and shape-memory properties was created at 25 bar/s depressurization rate in the temperature range between 78 and 84 °C. The shape-memory behavior of this foam was investigated using different programming modules, such as under stress-free condition and under constant strain condition. The thermally-induced shape-memory effect (SME) was found to be strongly dependent on the programming conditions. Excellent shape fixity has been observed for all foams indicating the high efficiency of the switching domains to fix the temporary shape by crystallization. The stress recovery of this foam could be controlled by changing compression percentage ($\varepsilon_c$%) at a constant compression temperature. The production of these foams with unprecedented properties by commercially available processing equipment raises much hope with the potential to provide new materials with a unique combination of shape-memory properties and porous structure as well as desired properties for many industrial and biomedical applications

INTRODUCTION

Shape-memory polymer (SMP) foams are unique, extraordinarily versatile materials that can be easily fabricated with very high porosity, low density, high compressibility, and shape-memory capability, i.e.; have the ability to memorize and return to its original permanent shape when exposed to an external stimulus. The external stimulus usually is heat (thermally-induced),[1,2] light,[3,4] magnetic field,[5] or electric field.[6] An SMP-foam is an example of a smart material with a wide range of potential applications. For example SMP-foams have been used in the area of aerospace applications, such as space deployable support structures, shelters for space habitation, and rover components, where the low density of the foams offset the lower mechanical properties compared to the un-foamed or bulk SMPs.[7,8] Foaming of SMPs is a technique, which can be used to tailor the properties for specific application requirements. Generally, mechanical strength and stiffness remarkably decrease by foaming, but the compressibility increases to a great extent. High pressure processing with supercritical carbon dioxide (scCO$_2$) is a well established technique for producing high porosity foams.[9] Porosity, pore size distribution, and foam density can be easily controlled by foaming temperature, depressurization rate, and saturation time. This technique is highly recommended for fabricating environmentally-friendly foams with no volatile organic solvents, suitable for biomedical applications. PDLCL is a class of multiblock copolymer comprising of poly($\omega$-pentadecalactone)

(PPDL) as a hard segment and poly($\varepsilon$-caprolactone) (PCL) as a soft segment. Both segments are linked by 2,2,4-trimethyl-1,6-hexamethylene diisocyanate (TMDI) as a junction unit through a co-condensation synthesis technique.[11] By varying molecular weight of the segments and the weight of switching to hard segments, a broad range of physical properties could be achieved.[11] The advantage of two crystallizable segments having different, distinct melting temperatures $T_m$ (PPDL) = 82 °C and $T_m$ (PCL) = 48 °C (as determined by DSC) allowed controlling the thermomechanical behavior and fabricating the material in different shapes as a function of temperature and mechanical history. The SME of these multiblock copolymers (bulk materials) of different switching segment contents using stretching thermomechanical tests has been previously investigated.[11] In the current study we explored, whether these multiblock copolymers can be processed into three-dimensional interconnected foam by scCO₂ foaming technique. It is expected that PPDL hard segment provides strong physical crosslinks ensuring sufficient mechanical stability of the foam especially at temperatures higher than $T_m$ of PCL. The shape-memory capability of these foams will be investigated by thermomechanical compression tests.

**EXPERIMENTAL DETAILS**

The PDLCL bulk material was synthesized from PPDL-diol, PCL-diol, TMDI and dibutyltin laurate as catalyst in 1,2-dichlorethane at 80 °C. The polymer was obtained by precipitation in methanol at -10 °C. More details about synthesis of this material is described elsewhere.[11,12] The obtained PDLCL bulk material has 60 wt% PCL content. The PDLCL has $M_n$ = 58,000 g/mol and 2.2 polydispersity based on GPC measurements.

Prior to the foaming process, the PDLCL bulk material was granulated at 95 °C in a single screw extruder (Thermo Electron l/d 34, d = 15 mm). Details of the foaming process using scCO₂ as a foaming agent is described elsewhere.[13] In this foaming process, the sample will be saturated with scCO₂ at elevated temperature ($T$ = 79 °C) and constant pressure (200 bar) for a given saturation time (30 min). The very quick depressurization rate (25 bar/s) leads to nucleation and expansion of dissolved CO₂, generating the foam. After discharging to zero overpressure the foams were removed from the vessel and stored under normal conditions over 24 h.

For SEM investigation the fracture cross-section of a foam was obtained using an edged tool after immersing the sample in liquid nitrogen for a few min. The cross-section foam was then fixed on holders with a conductive adhesive, and sputtered with magnetron (EMI Tec XY, Great Britain). The prepared samples were investigated using a LEO 1550 VP electron microscope with a Schottky emitter (LEO, Germany).

Three-dimension morphology of the foams was investigated using micro x-ray Computer Tomography, µCT Procon X-ray GmbH. The foam was fixed on a rotary stage and scanned inside the µCT apparatus. The morphology investigation was carried out at a voltage of 40 kV and a current of 0.19 mA. Porosity and pore size distribution can be also quantitatively determined from the 3D µCT image.

The shape-memory creation procedure, *SMCP*, is a three-steps thermomechanical conditioning process with the following steps:
(i) The foam will be heated to a high temperature $T_{high}$ = 70 °C and annealed at this temperature for equilibrium (10 min) to remove any thermal or mechanical history of the foam. (ii) The foam

100

will be cooled down to the programming temperature, $T_{prog}$ = 60 °C at 5 °C/min cooling rate and compressed to a constant value of $\varepsilon_c$ = 50, 65, or 80%. (iii) The foam will be cooled down (10 °C/min cooling rate) to $T_{low}$ = 0 °C under the applied compression and annealed for 10 min to fix the compressed shape. Finally, the foam will be heated to $T_{high}$ = 70 °C at 2 °C/min heating rate to return to its original uncompressed shape. The thermomechanical tests were performed on Zwick Z1.0 and Z005 tensile testers equipped with thermo chambers and temperature controller (Eurotherm Regler, Limburg, Germany). Foam cubes with a side of 12 mm were used for the thermomechanical-compression measurements. Two different programming modules, namely stress free condition and constant-strain conditions were employed for the thermomechanical studies.

## RESULTS AND DISCUSSION

The morphology of the foam was investigated by μCT and SEM as demonstrated in Figures 1a and 1b, respectively. The μCT is a non-destructive technique, which can be used for 3D visualization of biological material structures, eg. bone or polymer foams.[14] SEM photographs revealed a high degree of homogenous pores distribution. The degree of porosity reaches approximately ~89±4% as determined by pycnometry. According to the pycnometry measurements, 51±10% of the pores were accessible by nitrogen. The μCT investigation of the foam revealed an average pore size of 200±30 μm. Well-defined cell walls of 10-20 μm thick is determined from the SEM photograph. The average pore size calculated from SEM is approximately 170 μm in good agreement with that obtained using μCT.

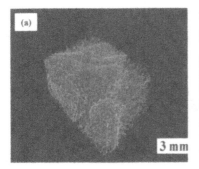
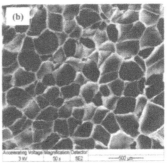

**Figure 1.** (a) Three-dimension μCT photograph for PDLCL foam. (b) SEM photograph for PDLCL foam. The foam was generated using scCO₂ foaming technique at 80 °C foaming temperature, 200 bar, and 25 bar/s depressurization rate.

The obtained foam has an excellent elasticity and can be easily programmed or compressed according to the shape-memory creation procedure (*SMCP*) mentioned above. To show a macroscopic active movement or compression-recovery behavior, the foam was compressed at $T_{prog}$ = 60 °C by $\varepsilon_c$ = 50% and then cooled to $T_{low}$ = 0 °C. The compressed foam was then heated in an electric oven at 2 °C/min heating rate. The thermally-induced recovery of this foam was recorded using a video camera. The compressed foam started to recover to its permanent uncompressed shape with increasing temperature as clearly seen in Figure 2. The foam reached a recovery rate of approximately 85% at $T$ = 70 °C.

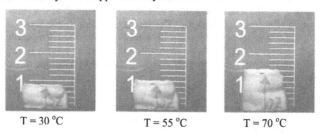

T = 30 °C          T = 55 °C          T = 70 °C

**Figure 2.** Macroscopic recovery process at different temperatures of a foam temporally fixed in a compressed form. The foam was programmed at $T_{prog}$ = 60 °C, $\varepsilon_c$ = 50%, and $T_{low}$ = 0 °C. The recovery took place by heating the foam in an electric oven at 2 °C/min heating rate.

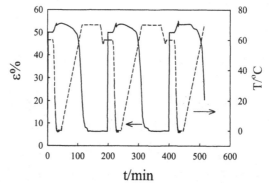

t/min

**Figure 3.** Cyclic, thermomechanical compression tests carried out at $T_{prog}$= 60 °C for different cycles. Before each cycle the foam was heated up to $T_{high}$ = 70 °C to remove any mechanical and thermal history particularly for the PCL segments.

A typical example for a cyclic thermomechanical compression test under compressive stress-control condition for three cycles under identical *SMCP* condition ($T_{high}$ = 70 °C, $T_{prog}$ = 60 °C, $\varepsilon_c$ = 50%, and $T_{low}$ = 0 °C) is presented in Figure 3. Here compression and sample temperature as a function of time are displayed. The sample was compressed at 60 °C by $\varepsilon_c$ = 50% and then cooled down quickly to $T_{low}$ = 0 °C. Obviously, an additional compression (about 4%) was observed during the cooling step from 60 to 0 °C under the constant compression (50%) for all cycles. This additional increase in the compression is attributed to the very fast crystallization process of PCL (switching domains) which leads to a highly ordered and compact structure and consequently induced a contraction or compression besides to the initial value. It might also be attributed to compression-induced crystallization of this foam.[15-17] The compressed foam in its temporary shape started to recover when heated. In the first cycle, the foam maintained in its compressed temporary shape for about 90 min even after increasing the temperature up to about 40 °C at 2 °C/min heating rate. At higher temperatures the sample recovered rapidly and reached about 85% recovery rate ($R_r$) for the first cycle within approximately 20 min by increasing the temperature up to 70 °C. $R_r$ reached approximately 98% for the second and third cycles. The recovery of the original shape is attributed to the fact that with increasing temperature the foam can regain the entropy, which was lost during the compression step. The shape fixity rate, $R_f$ calculated from the value of $\varepsilon_u$ (obtained directly after release the external compression) divided by $\varepsilon_c$ was found to be independent on the cycle number ($R_f$ is 100% for all cycles). The excellent shape fixity reflects the high efficiency of the switching segment to fix the temporary shape by crystallization. The switching temperature $T_{sw}$ was also calculated for each cycle. $T_{sw}$ was found to be constant regardless of the cycle number. $T_{sw}$ is evaluated by following the change in recovery during the transition from the compressed-temporary shape to the uncompressed permanent shape. The value of $T_{sw}$ was 67 °C for all cycles carried out at $T_{prog} = 60$ °C.

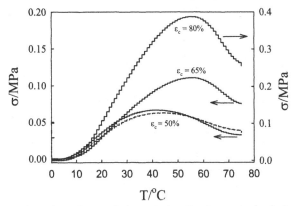

**Figure 4.** Temperature dependence of stress during the thermomechanical-compression test (strain control) for compressed foam under different $\varepsilon_c\%$ ($\varepsilon_c$ = 50, 65, 80, and 50%).

Thermomechnical tests under constant compressive strain control conditions were employed to measure the recovery stress of this foam. In this case the sample was programmed under a typical *SMCP* condition as described above. When the compressed foam was reheated under a fixed compressive strain condition, the foam generated a strong stress peak ($\sigma_{max}$) with a maximum at a well-defined temperature ($T_{\sigma,max}$). In this experiment, the recovery stress built up with increasing temperature until the softening of the foam allows stress relaxation. This peak was also observed for other SMPs under fixed strain condition.[18,19] Figure 4 shows how the value of $\varepsilon_c$ could be used to obtain different values of $\sigma_{max}$ and $T_{\sigma,ma}$. $\sigma_{max}$ and $T_{\sigma,max}$ significantly increased with increasing $\varepsilon_c$ as clearly seen in the Figure. For these measurments, four cycles of the thermomechnical tests were carried out at different $\varepsilon_c$ = 50, 65, 80 and 50%. The first and fourth cycles at $\varepsilon_c$ = 50% generated almost the same $\sigma_{max}$ and $T_{\sigma,max}$ (see Figure 4). This experimental fact suggested that the foam could keep its high stress recovery rate even after four different cycles with thermal and compression treatments.

## CONCLUSIONS

Based on the preceding discussion, it can be concluded that foaming of PDLCL biodegradable multiblock copolymers with scCO$_2$ resulted in highly porous materials with a unique morphology in a short saturation time. The shape-memory effect of this foam has been quantitatively investigated under stress-free condition and under constant strain condition ($\varepsilon_c$ = constant). The production of this SMP-foam by commercially available processing equipment resulted in a unique combination of porous structure and excellent shape-memory properties for many industrial and biomedical applications, such as aerospace and tissue engineering. The present study may stimulate a better understanding of the rational design of SMP-foams with unique and interesting properties.

## REFERENCES

1. M. Behl, A. Lendlein, *Soft Matter*, **2007**, *3*, 58
2. Bellin, S. Kelch, R. Langer, A. Lendlein, *Proc. Natl. Acad. Sci. USA*, **2006**, *103*, 18043-18047.
3. H. Jiang, S. Kelch, A. Lendlein, *Adv. Mat.*, **2006**, *18*, 1471-1475.
4. A. Lendlein, H. Jiang, O. Jünger, R.Langer, Nature, **2005**, *434*, 879-882.
5. R. Mohr, K. Kratz, T. Weigel, M. Lucka-Gabor, M. Moneke, A. Lendlein, *Proc. Natl. Acad. Sci. USA*, **2006**, *103*, 3540-3545.
6. J. Leng, H. Lv, Y. Liu, S. Du, *J. Appl. Phys.* **2008**,*104*, 104917.
7. H. Tobushi, K. Okumura, M. Endo, S. Hayashi, *J. Intell. Mater. Syst. Struct.*, **2001**, *12*, 283.
8. M. Sokolowski, S. Hayashi, T. Yamada, "Cold Hibernated elastic memory (CHEM) self deployable structures Smart Structures and Materials: Electroactive Polymer Actuators and Devices **1999** (Bellingham, WA: SPIE Optical Engineering Press).
9. H. Tai, M. L. Mather, D. Howard, W. Wang, L. J. White, J. A. Crowe, S. P. Morgan, A. Chandra, D. J. Williams, S. M. Howdle, K. M. Shakesheff, *Eur. Cell. Mater. J.*, **2007**, *14*, 64.
10. P. Ping, W. S. Wang, X. S. Chen, X. B. Jing, *Biomacromolecules* **2005**, *6*, 587.
11. K. Kratz, U. Voigt, W. Wagermaier, A. Lendlein, in Advances in Material Design for Regenerative Medicine, Drug Delivery, and Targeting/Imaging, Material Research Society Symposium Proceedings Volume 1140, Warrendale, PA, 2009), 1140-HH03-01
12. A. Kumar, B. Kalra, A. Dekhterman, R. Gross, *Macromolecules* **2000**, *33*, 6303-6309.
13. K. Luetzow, F. Klein, T. Weigel, R. Apostel, A. Weiss, A. Lendlein, *J. Biomech.* **2007**, *40*, S80.
14. R. Filmon, N. Retailleau-Gaborit, F. Grizon, M. Galloyer, C. Cincu, M. F. Basle, D. Chappard, *J. Biomater. Sci. Polym. Ed.* **2002**, *13*, 1105-1117.
15. K. Kojio, S. Nakamura, M. Furukawa, *Polymer* **2004**, *45*, 8147-8152.
16. H. Koerner, G. Price, N. A. Pearce, M. Alexander, R. A. Vaia, *Nature Materials* **2004**, 3, 115.
17. M. Okamoto, H. Kubo, T. Kotaka, *Macromolecules* **1999**, 32, 6206-6214.
18. K. Gall, C. M. Yakacki, Y. P. Liu, R. Shandas, N. Willett, K. S. Anseth, *Journal of Biomedical Materials Research, Part A* **2005**, *73A*, 339.
19. Y.Miyamoto, K. Fukao, H. Yamao, K. Sekimoto, *Phys. Rev. Lett.* **2002**, *88*, 225504

Mater. Res. Soc. Symp. Proc. Vol. 1190 © 2009 Materials Research Society                    1190-NN06-02

# Thermoemechanical Properties and Shape-Memory Capability of Drug Loaded Semi-Crystalline Polyestermethacrylate Networks

Axel T. Neffe[1,2,] Bui D. Hanh[1], Susi Steuer[3], Christian Wischke[1,2] and Andreas Lendlein[1,2]
[1]Centre for Biomaterial Development, GKSS Forschungszentrum Geesthacht, Kantstrasse 55, 14513 Teltow, Germany
[2]Berlin-Brandenburg Centre for Regenerative Therapies (BCRT), Berlin, Germany
[3]Present address: Intervet Innovation GmbH, Zur Propstei, 55270 Schwabenheim, Germany

## ABSTRACT

Polymer networks synthesized by UV-curing of Oligo[(ε-caprolactone)-*co*-glycolide]-dimethacrylates are hydolytically degradable. Their architecture with covalent netpoints and crystallizable domains is the molecular basis for a potential shape-memory capability. The molecular weight and glycolide content of the oligomeric precursors can be varied over a broad range of compositions to tailor the thermomechanical properties of the polymer network while having only a minor influence on the shape-memory effect. Recently, drug incorporation adding controlled drug release as further functionality to the polymer network was demonstrated [4]. Here, enoxacin and ethacridine lactate as test drugs were incorporated into the networks by soaking. Alternatively, defined amounts of ethacridine lactate were mixed with the precursors, which were subsequently crosslinked to the drug containing networks. The composition of the oligomeric precursors was varied in molecular weight between 3800 and 12800 $g \cdot mol^{-1}$ and in glycolide content $\chi_G$ between 0 and 30 mol-% to explore the influence of the drug incorporation on networks with varying compositions while retaining properties and functionalities. Polymer networks prepared from precursors with $\chi_G \leq 14$ mol-% and $M_n \geq 6900$ $g \cdot mol^{-1}$ have a $T_{sw}$ of 35-52 °C and sufficient crystallinity to ensure a high shape fixity in the programming step. These limits have to be kept to ensure the desired multifunctionality, otherwise drug incorporation can have an undesired influence on thermal, mechanical, and shape-memory properties.

## INTRODUCTION

Polymer systems are families of polymers in which slight changes of structural parameters allow the adjustment of the overall bulk properties of the material in a wide range [1]. This behavior can be used to tailor the desired properties to the specific application. Functionalities of polymers are different from properties as they are non-inherent, but are incorporated by a specific process. A typical example is the shape-memory effect; while being dependent on the molecular architecture/morphology of the polymer, a programming step is necessary to fix the deformation of a polymer as a temporary shape. Only an external trigger such as increased temperature leads to the recovery of the permanent shape [2]. Possible applications of shape-memory polymers include minimally invasive surgery [2] and intelligent suture material [3].

Biomaterials need to fulfill complex requirements, for which a combination of several functions has to be combined with tailored properties. Therefore, multifunctional materials are needed, in which the processing steps for incorporating the functionalities have to be compatible with each other, and it has to be explored, in which limits the parameters deter-mining the properties of the polymer can be varied without affecting the functionalities of the polymer. Recently, it has been demonstrated that it is possible to combine the three function-alities biodegradability, shape-memory capability, and controlled drug release in one polymer based material [4]. For this purpose, polymer networks were chosen, which were synthesized by UV-curing of oligo[(ε-caprolactone)-co-glycolide]-dimethacrylates. The temporary shape of these polymer networks was fixed by crystallization of the oligo(ε-caprolactone) segments [5]. Here, the influence of drug loading on the thermomechanical and shape-memory is explored for networks with varying molecular weight and glycolide content of the precursors.

## EXPERIMENTAL DETAILS

### Synthesis and thermomechanical characterization

The synthesis of the oligo[(ε-caprolactone)-co-glycolide]-dimethacrylates and polymer network formation are described in Ref. 5. The oligomers were synthesized by dibutyltin(II) oxid catalyzed ring-opening polymerization of diglycolide and ε-caprolactone using ethylene glycol as starter molecule, methacrylation of the terminal OH-groups, and subsequent crosslinking by UV-curing.

Thermal properties were determined by Differential Scanning Calorimetry (DSC), which was performed on a Perkin-Elmer DSC7. Tensile tests were performed at 37 °C on dumbbell-shaped specimen with a gauge length of 3.0 mm and a total length of 10.0 mm on a Zwick 1425 (Zwick GmbH) with a Climatix thermochamber and a Eurotherm temperature controller. The samples were immersed in a 2 L water bath which was connected to a thermostat Ecoline (RE 106, Lauda).

Cyclic, thermomechanical tests were performed on specimen with dimensions as described above for tensile tests. As tensile testing machine, a Zwick 2.5N1S, (Zwick GmbH & Co.) equipped with a 50 N load cell was used. The program for strain-controlled cyclic, thermomechanical tests consisted of stretching the sample to maximum extension $\varepsilon_m = 75\%$ with $T_{high} = 70$ °C, and $T_{low} = 0$ °C. To precondition the specimen for the stresscontrolled cyclic, thermomechanical experiments, the sample was heated to $T_{high} = 70$ °C for 5 min and strained to $\varepsilon_m = 100\%$ for three times. Stress-controlled test cycles were performed with $\varepsilon_m = 100\%$, $T_{high} = 70$ °C, and $T_{low} = 15$ °C.

### Drug loading, drug release, and hydrolysis experiments

Drug Loading was achieved either by swelling of the respective polymer network in a 100 fold excess (V/m) of the polymer in a saturated solution of enoxacin (EN) in chloroform or of ethacridine lactate (EL) in a 1:1 mixture (V/V) of chloroform and 2-propanol, respectively, at room temperature for 24 h. This time was chosen as no further drug loading was observed after this time (data not shown). Alternatively, defined amounts EL were added to a 10% (m/V) solution of the macrodimethacrylates in a 1:1 (m/m) mixture of dichloromethane and 2-propanol. The mixture was concentrated at 50 °C and subsequently completely dried at 70 °C under reduced pressure (1 mbar) for 2 h. This mixture was subsequently used for the crosslinking. Drug loading of the networks was determined by complete acidic methanolysis of the networks and subsequent determination of drug concentration by UV spectroscopy [4].
Hydrolysis experiments and drug release were performed in aqueous phosphate buffer pH 7.0 at 37 °C [4]. Drug concentration in the medium was determined by UV spectroscopy. Three test pieces of each polymer were analyzed.

## RESULTS AND DISCUSSION

The number average molecular weight $M_n$ of the precursor molecules was varied between 3800 g·mol$^{-1}$ and 12800 g·mol$^{-1}$, and the glycolide molar fraction $\chi_g$ between 0 and 30 wt.-% (Nomenclature for the networks used in this paper is N-CG($\chi_G$)-$M_n$ with $\chi_G$ in wt.-% and $M_n$ in $10^3$ g·mol$^{-1}$). EL and EN were incorporated as test drugs into the polymer networks either by swelling of the networks in concentrated solutions of the drug and subsequent drying (final loading: 0.72 wt.-% EN or 0.60 wt. % EL) or alternatively by dispersion of the drugs in the precursors prior to the crosslinking (*in situ* drug incorporation, 0.2-5.7 wt.-% EL). The test drugs were chosen for their different physicochemical properties (EL is much more hydrophilic than EN).

### Thermal Properties

Thermal properties of unloaded and drug loaded networks were determined by DSC. The unloaded and drug loaded networks showed a glass transition $T_g$ of -59 to -52 °C and a melting temperature $T_m$ of 19-55 °C (Full data is given in Table 1). As was observed for the unloaded networks [5], $T_g$ of the loaded networks is nearly constant over a wide range of sample compositions, while $T_m$, which is related to the crystalline poly($\epsilon$-caprolactone) (PCL) domains, increased with increasing $M_n$ of the precursors, and decreased with increasing glycolide content. Drug loading by swelling did not show strong effects on either $T_g$ or $T_m$. A typical profile for the $T_m$ of drug free and drug loaded networks with different compositions is depicted in Figure 1a. The melting enthalpy $\Delta H_m$ decreases by up to 4 J·g$^{-1}$ when drug is soaked into networks with already low crystallinity, i.e. with either a high glycolide content or a low $M_n$ of the precursors such as the networks N-CG(30)-10 and N-CG(14)-3. This can be attributed to a further disturbance of the crystallization of the polymer chains by drug molecules. Increasing drug loading by the *in situ* process did not have a unidirectional influence on $T_m$; in fact, networks with 2-4.8 wt.-% did have slightly higher $T_m$s (42 °C) than networks with lower or higher drug loading (ca. 38 °C). $\Delta H_m$ decreases with increasing drug content (Fig. 1b), which suggests a negative influence of the drug molecules on the crystallization behavior of the polymer matrix.

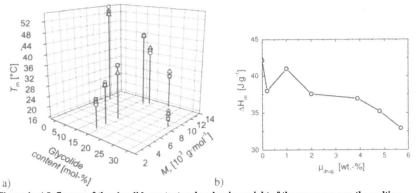

a)                                                        b)

**Figure 1:** a) **Influence of the glycolide content and molecular weight of the precursors on the melting temperature of unloaded networks and networks loaded by swelling. △ unloaded N-CG network, ○ EN loaded N-CG network, □ EL loaded N-CG network. b) Melting enthalpy of the polymer network N-CG(14)-10 depending on the drug content. (Values have been scaled to the mass of the matrix only.)**

## Mechanical Properties

Mechanical properties of the materials were measured under physiological conditions (i.e. at 37 °C after equilibration of the samples in aqueous medium), as these are conditions for the application of the networks described. The tensile tests depicted in Figure 2 were performed with the networks N-CG(0)-10, N-CG(14)-10, and N-CG(21)-10, either unloaded or with incorporation of 1 wt.-% EL *in situ*. Network N-CG(0)-10 has a $T_m$ of 53 °C and is semi-crystalline under the conditions of the experiment. Young's modulus E and tensile strength $\sigma_{max}$ are very similar for the loaded and unloaded networks, however the unloaded network has a higher elongation at break $\varepsilon_b$. The other networks have $T_m$s lower than 37 °C and are therefore amorphous under the conditions of the experiment. In this case, drug loading has a more pronounced effect on the mechanical properties, which is e.g. demonstrated by the reduction of Young's modulus of N-CG(21)-10 of 3.6 MPa to 0.8 MPa by the drug loading. Drug molecules potentially act as plasticizer here. The trend observed for the thermal properties, i.e. stronger influence of drug loading on networks with low crystallinity, can likewise be observed for the mechanical properties. Mechanical properties of networks determined at 22 °C were not influenced by drug loading up to 4.8 wt.-%.

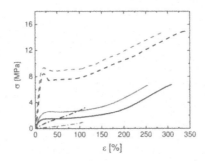

**Figure 2: Tensile tests of the networks ---N-CG(0)-10, —N-CG(14)-10, and -.-.-N-CG(21)-10 at 37 °C after equilibrating the samples in water. Grey lines represent networks loaded with 1 wt.-% EL.**

## Shape-Memory Functionality

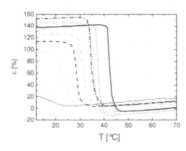

**Figure 3: Influence of drug loading on the shape recovery of the polymer networks in a stress controlled thermomechanical experiment with variation of the $M_n$ of the precursors. ----N-CG(14)-3; ----N-CG(14)-3-EL; --- N-CG(14)-5; ---N-CG(14)-5-EL; -.-.-N-CG(14)-7; -.-.-N-CG(14)-7-EL; — N-CG(14)-10; —N-CG(14)-10-EL.**

108

Shape-memory properties were quantified by cyclic, thermomechanical tensile tests [2,6]. Stress- and strain-controlled experiments were employed to the unloaded and drug loaded polymer networks in air. Strain fixity $R_f > 94\%$ and strain recovery $R_r > 97\%$ were observed for polymers with a precursor glycolide content of up to 14 mol % and $M_n$ of the macrodimethacrylates of at least 6900 $g \cdot mol^{-1}$ (Table 1), also when loaded by swelling. The shape-memory effect for these polymer networks is not influenced by drug loading, and a quantitative shape recovery was obtained with only a small change in $T_{sw}$. As $R_f$ is ruled by crystallizable switching segments, the influence of drug loading on the N-CG networks with lower $M_n$ of the precursors or higher glycolide content, which show only a small degree of crystallinity, is consequently increased. Figure 3 contains the recovery curves from stress-controlled tests of loaded (swelling) and unloaded N-CG networks under variation of the precursor $M_n$ visualizing these findings. Most importantly, drug incorporation decreases $R_f$.

**Table 1:** Thermal properties and quantified shape-memory effects of the networks depending on $M_n$ and glycolide content of the precursors and the drug loading by swelling. $T_g$ = glass transition temperature, $\Delta C_P$ = change of heat capacity at the glass temperature, $T_m$ = melting temperature, $\Delta H_m$ = melting enthalpy, $\varepsilon_1$ = elongation after cooling in the programming procedure, $R_f$ = strain fixity, $R_r$ = strain recovery, $T_{sw}$ = switching temperature, [a] = determined in water. n.d. = not determined.

| Polymer network | $T_g$ °C | $\Delta C_P$ $J \cdot (g \cdot K)^{-1}$ | $T_m$ °C | $\Delta H_m$ $J \cdot g^{-1}$ | $\varepsilon_1$ % | $R_f$ % | $R_r$ % | $T_{sw}$ °C |
|---|---|---|---|---|---|---|---|---|
| N-CG(14)-3 | -56 | 0,37 | 26 | 32 | 123 | 80,6 | 100 | 23 |
| N-CG(14)-3-EN | -55 | 0,48 | 24 | 28 | n.d. | n.d. | n.d. | n.d. |
| N-CG(14)-3-EL | -55 | 0,49 | 25 | 29 | 119 | 23,4 | 96,7 | 19 |
| N-CG(14)-5 | -53 | 0,28 | 30 | 36 | 147 | 86,3 | 96,9 | 31 |
| N-CG(14)-5-EN | -53 | 0,23 | 31 | 36 | n.d. | n.d. | n.d. | n.d. |
| N-CG(14)-5-EL | -52 | 0,23 | 30 | 36 | 151 | 76,1 | 98,2 | 28 |
| N-CG(14)-7 | -53 | 0,27 | 36 | 41 | 145 | 98,4 | 97,5 | 36 |
| N-CG(14)-7-EN | -53 | 0,25 | 36 | 39 | n.d. | n.d. | n.d. | n.d. |
| N-CG(14)-7-EL | -52 | 0,29 | 33 | 37 | 129 | 98,7 | 97,9 | 35 |
| N-CG(14)-10 | -53 | 0,21 | 38 | 42 | n.d. | 99,8[a] | 95,1[a] | 35[a] |
| N-CG(14)-10-EN | -53 | 0,27 | 39 | 42 | n.d. | 99,6[a] | 96,9[a] | 35[a] |
| N-CG(14)-10-EL | -54 | 0,28 | 39 | 42 | 142 | 97,2 | 100 | 42 |
| N-CG(0)-10 | -59 | 0,2 | 53 | 59 | 151 | 91,9 | 100 | 52 |
| N-CG(0)-10-EL | n.d. | n.d. | 55 | 59 | 136 | 94,5 | 100 | 52 |
| N-CG(21)-10 | -53 | 0,3 | 28 | 36 | 143 | 68,8 | 100 | 27 |
| N-CG(21)-10-EL | n.d. | n.d. | 27 | 34 | n.d. | n.d. | n.d. | n.d. |
| N-CG(30)-10 | n.d. | n.d. | 20 | 19 | n.d. | n.d. | n.d. | n.d. |
| N-CG(30)-10-EN | n.d. | n.d. | 19 | 16 | n.d. | n.d. | n.d. | n.d. |
| N-CG(30)-10-EL | n.d. | n.d. | 19 | 15 | n.d. | n.d. | n.d. | n.d. |

## Further Functionalities of the Networks

The hydrolytic cleavage of the ester bonds is the reason for the degradability of the polymer networks, however, the methacrylate polymer chains formed in the photocrosslinking step are obviously not degraded by hydrolysis. The degradation rate is decreasing slightly with increasing $M_n$ of the precursors, while an increasing glycolide content of the precursors increased the degradation rate more prominently. Drug loading did not influence the degradation rate (Data not shown). Drug release is diffusion controlled [4]. Figure 4 depicts the drug release depending on the $M_n$ of the precursors (a) and on the glycolide content (b).

Generally, the more hydrophilic EL is released faster than the more hydrophobic EN, drug release is decreasing with increasing glycolide content, while no clear dependence of release rate and chain segment length has been found.

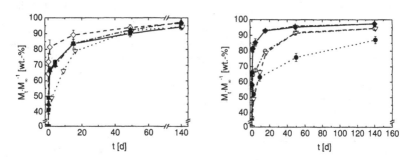

**Figure 4: Release of ethacridine lactate from networks under variation of the chain segment length (a) and the glycolide content (b). ■ N-CG(14)-3-EL; ▲N-CG(14)-5-EL; ◇ N-CG(14)-7-EL; ▽ N-CG(14)-10-EL. ● N-CG(0)-10-EL; ○ N-CG(12)-10-EL; ★ N-CG(21)-10-EL;◆ N-CG(30)-10-EL.**

## CONCLUSIONS

It has been demonstrated, that drug incorporation into the networks has an influence on the thermomechanical properties and shape-memory functionality, if the polymer network is of low crystallinity (i.e. the precursor having a low $M_n$ or a high $\chi_G$) or if high amounts of drug ($\geq$ 5.7 wt.-%) are incorporated. Such high drug loadings were reached in this study only by dispersion of the drug in the prepolymers with subsequent crosslinking. Drug incorporation itself has a negative influence on the crystallization behaviour of the polymer matrix, which is represented by lower melting enthalpy, lower Young's modulus, and lower strain fixity of the networks, while physicochemical properties of the drug such as hydrophilicity are of no concern. When choosing networks prepared from precursors with $\chi_G$ $\leq$ 14 mol-% and $M_n \geq 6900$ g·mol$^{-1}$, the networks have a high enough tendency to form crystalline phases, so it is possible to tailor the thermomechanical properties of the loaded networks as it has been demonstrated for the unloaded networks before. Such networks furthermore have a $T_{sw}$ close to or above body temperature, an adjustable hydrolytic degradation rate, a high strain fixity in the programming step, and show diffusion controlled release of incorporated drugs, which results in multifunctional materials. It will be of especial interest to increase drug loading by measures other than the *in situ* incorporation such as by employing a polymer network with a further amorphous, advantageously hydrophobic, phase and/or different network architecture. The polymer networks in this study were bulk degrading, a potential surface degrading material could furthermore broaden the scope of application.

## ACKNOWLEDGMENTS

The authors thank for financial support by the BMBF Biofuture grant Nr. 0311867.

## REFERENCES

1. A. Lendlein, S. Kelch. *Materials Science Forum* **2005**, *492-493*, 219-223.
2. M. Behl, A. Lendlein, *Soft Matter* **2007**, *3*, 58-67.
3. A. Lendlein, R. Langer. *Science* **2002**, *296*, 1673 – 1676.
4. A.T. Neffe, B.D. Hanh, S. Steuer, A. Lendlein, *Adv. Mat.* **2009**, *21*, 3394–3398.
5. S. Kelch, S. Steuer, A.M. Schmidt, A. Lendlein. *Biomacromolecules* **2007**, *8*, 1018-1027.
6. R. Mohr, K. Kratz, T. Weigel, M. Lucka-Gabor, M. Moneke, A. Lendlein, A. *Proc. Natl. Acad. Sci. U.S.A.* **2006**, *103*, 3540-3545.
7. B. Narasimhan, R. Langer, *Polym. Mater. Sci. Eng.* **1997**, *76*, 558-559.
8. R. Wada, S.H. Hyon, T. Nakmura, Y. Ikada, *Pharm. Res* **1991**, *8,* 1292-1296.

Mater. Res. Soc. Symp. Proc. Vol. 1190 © 2009 Materials Research Society          1190-NN06-05

# In vivo degradation behavior of PDC multiblock copolymers containing poly(p-dioxanone) hard segments and crystallizable poly(ε-caprolactone) switching segments

B. Hiebl[1,2], K. Kratz[1,2], R. Fuhrmann[3], F. Jung[1,2], A. Lendlein[1,2], R-P. Franke[1,2,3]
[1] Center for Biomaterial Development, Institute of Polymer Research, GKSS Research Center, Kantstrasse 55, D-14513 Teltow, Germany
[2] Berlin-Brandenburg-Center for Regenerative Therapies, Charité- Universitätsmedizin Berlin_ BCRT, Campus Virchow-Klinikum, Augustenburger Platz 1, D-13353 Berlin, Germany
[3] University of Ulm, Central Institute for Biomedical Engineering, Department of Biomaterials, Ulm, Germany

## ABSTRACT

The degradation behavior of biodegradable multiblock copolymers (PDC) containing poly(p-dioxanone) hard segments (PPDO) and crystallizable poly(ε-caprolactone) switching segments (PCL) synthesized via co-condensation of two oligomeric macrodiols with an aliphatic diisocyanate as junction unit was explored in *in vivo* and *in vitro* experiments. The *in vitro* experiments for enzymatic degradation resulted that the poly(ε-caprolactone) segments are degraded faster, than the poly(p-dioxanone) segments. During degradation the outer layer of the test specimen becomes porous. Finally non-soluble degradation products in form of particles were found at the surface. This observation is in good agreement with the in vivo studies, where the non-soluble degradation products in the periimplantary tissues showed a diameter of 1 – 3 micron.

## INTRODUCTION

For medical applications of hydrolytically degradable biomaterials like surgical implants, e.g. sutures, or implantable drug depots e.g. for treatment of cancer, a detailed knowledge of the degradation behaviour during use in patient is required. A major limitation of established degradable biomaterials like poly(α-hydroxy esters) is the fact, that their degradation behaviour during clinical application can not be predicted reliably based on the results of *in vitro* experiments [1]. Therefore more fundamental research is required to enable a knowledge based approach for biomaterial development. Multiblock copolymers with shape-memory capability attracted tremendous interest as promising candidates for enabling new medical applications like an intelligent surgical suture [2-6]. In the present work we investigated the degradation behavior of PDC multiblock copolymers, containing poly(p-dioxanone) hard segments (PPDO) and crystallizable poly(ε-caprolactone) switching segments (PCL). These copolyesterurethanes were synthetised by co-condensation of two oligomeric macrodiols with an aliphatic diisocyanate [4,6]. The PDC multiblock copolymers exhibited a linear mass loss during hydrolytic degradation experiments, whereby the rate of the mass loss was dependent on the PPDO content [4]. In PCL selective enzymatic degradation studies using Pseudomonas lipase [7,8] selectivity of the enzyme towards the degradation of the PCL-segments lead to increased degradation rates in case of higher PCL-contents. In this study, the in vivo degradation behavior of PDC was investigated and compared to the results of the in vitro degradation in presence of enzymes. Emphasis was put on exploring the non-soluble degradation products.

## MATERIALS AND METHODS

### Multiblock Copolymer

The synthesis of PDC multiblock copolymers was previously described elsewhere [4]. For *in vivo* tests we selected a PDC multiblock copolymer with 60 wt-% PCL content. Disc-shaped test specimen with a diameter of 12mm and a thickness of 0.57mm were used for implantation.

### Surgical procedure – in vivo study

The study was licensed by the regional board of Giessen (Germany). Experiments were performed under specific pathogen-free conditions in the animal facility of the Phillips-University Marburg (Marburg, Germany) on adult Sprague-Dawley rats (310-360 g). Animals were maintained in isolated ventilated cages (one animal per cage). For implantation they were premedicated with atropine (0.05 mg/kg) and anesthetized with a mixture of ketamine (75 mg/kg) and xylazine (12 mg/kg). An incision of 1.5 cm was made dorsomedial between the scapulas. Connective tissue of the subcutis was prepared to house the implant. The copolymer-disc was implanted in 3 rats (group A), 3 animals were sham operated (group B). After surgery wound closure was done by suture (3-O Vicryl, Ethicon).

### Histology

Directly after explantation each polymer-tissue sample was cryoconserved. Briefly a drop of Tissue Teck® was prepared on a cork plate and the samples were aligned within this embedding medium. Thereafter the cork plate with the sample on top was placed on liquid nitrogen until the whole sample was frozen. The frozen samples were stored at -80 °C and fixed with methanol/ethanol (1:1, v:v) before sectioning and staining. Sections (thickness: 5 μm) were made using cryo-sectioning (Leica CM 3050S). Immunohistochemistry was performed to visualize laminin (primary antibody: polyclonal rabbit anti-laminin, 1:20; secondary antibody: polyclonal donkey anti-rabbit IgG, conjugated with TRITC (red), 1:50) and fibronectin (primary antibody: monoclonal mouse anti-fibronectin, 1:20; secondary antibody: polyclonal goat anti-donkey IgG, conjugated with FITC (green), 1:50). From every sample five sections were made, each section was evaluated at five different fields of view with a confocal laser scanning microscope (Leica LSM 5).

### Enzymatic degradation procedure

For the enzymatic degradation studies polymer film samples (10 x 10 mm) with a thickness of 0.5 mm were used. Each specimen was placed in a vial filled with 5 ml, pH 7.2 phosphate buffer (0.05 M) containing 200 μg·ml$^{-1}$ amount of Pseudomonas cepacia lipase. The vials were placed in water bath thermostated at 37 °C. The buffer/enzyme system was exchanged every 24 h to restore the original level of enzymatic activity. Degradation studies were performed with samples immersed in the degrading medium for 24 h, 48 h, 72 h and 100 h. After immersion in the degrading medium, the specimens were removed, washed with distilled water

(Milli-Q) and vacuum-dried at room temperature for one week before being subjected to analysis.

## Scanning electron microscopy

For scanning electron microscopy (SEM) investigations the dried samples were cut in small rectangular pieces and fixed on SEM sample holders. After gold–palladium coating within a cool sputter coater SCD 050 (Bal-tec, SEM Liechtenstein) the surface structure was studied in a high resolution scanning electron microscope LEO 1550 (Carl Zeiss, Oberkochen, Germany) at an accelerating voltage of 3 kV under high vacuum conditions (p ~ $10^{-6}$ Torr).

For the cross-section experiments, the sample pieces were cut with a diamond knife of an ultra-microtome (UCT, Leica, Germany).

## RESULTS AND DISCUSSION

As a result of the in *vivo study* table 1 gives the mean number of non-soluble degradation products in the tissues directly adjacent to the PDC disc three weeks after implantation. Most degradation products had a diameter lower than 10 µm. The number of degradation products with a diameter of more than 30 µm was significantly higher (p<0.0001) in the animals with the PDC implant in the neck as in the sham animals. In the sham animals no particles with a diameter of more the 1µm could be detected. Later particles might be incorporated by surgical instruments which were used also for the explantation of the copolymer implants.

Table 1:   *In vivo* study; non-soluble degradation products in the connective tissue adjacent to a PDC polymer disc (Ø 12 mm, thickness 0.57±0.02 mm) three weeks after implantation in the subcutaneous tissue of a rat neck; means ± standard deviation, n=75.

| Diameter of the degradation products [µm] | PDC | Sham |
|---|---|---|
| >30 | 3±5 | 0 |
| <10 | 19±11 | 0 |
| < 1 | 23±3 | 2±3 |

Figure 1A shows the high number of irregular round shaped PDC non-soluble degradation products with as size between 1 and 50 µm as yellow-green stained particles at the PDC-tissue-interface. These particles are neighboured by curled fibronectin (intense green) and laminin structures (weak green) of the connective tissue. Figure 1B shows the low particle load of the prepared implant sites in the connective tissue of the neck from the sham animals.

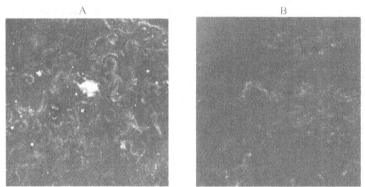

**Figure 1:** A: Periimplantary region of a PDC disc (Ø 12mm, thickness 0.57±0.02mm) three weeks after implantation in the subcutaneous tissue of a rat neck (snapshot, transillumination, magnification 1:630); B: Periimplantary region of the implant site in the neck subcutaneous tissue from the sham rats three weeks after surgery (snapshot, transillumination, magnification 1:630)

## DISCUSSION

For degradable polymer-based biomaterials like poly-[(*L*-lactide)-*co*-glycolide] it is supposed that the integration of those polymers into the surrounding tissues in terms of a strong connection between the implant and the periimplantary tissue is strongly dependent on the clearance of the degradation products arising at the polymer-tissue interface [9]. The *in vivo* study firstly presents a reliable method enabling the visualization and quantification of the non-soluble degradation product content ex vivo. Limitations are given by the fixation technique used to stabilise the sample structures and also by the sample size because of the limited diffusion distance of antibodies in tissues. The results confirm with PDC, that polymers can be able to bind fibronectin and laminin on their surface [10]. This material immanent feature enables to quantify the content of degradation products with an average diameter of about 1μm in the periimplantary tissues using a confocal laser-scanning microscope. The fluorescence signal of degradation products decreases with the particle size. Because of this the method is limited to degradation products with a diameter of about 1μm.

Most particulate degradation products had a diameter lower than 10 μm, whereas the majority of these particles showed diameters of round about 1 to 3 μm (see table 1). This is in good agreement with results recently described by Kulkarni et al. from an *in vitro* study [7]. In this study, the morphological changes on the surface induced by the enzymatic degradation process was monitored by scanning electron microscopy investigations. In Fig. 2 the SEM micrographs of PDC with 67 % PCL content degraded in phosphate buffered (pH 7.2) with lipase concentration of 200 μg·ml⁻¹ for 24 h, 72 h and 100 h clearly show a tremendous change of the surface roughness/morphology with degradation time. After 24 h enzymatic degradation the surface got strongly roughened and a crater landscape was formed with some grainy structures at the surface. After 72 h and 100 h degradation the surface roughness increased because of enhanced erosion, in addition to that micron-sized particles are visible on the polymer surface. Most of such microparticles have a nano-scaled porous substructure as displayed in Fig. 2 (100 h).

**Figure 2:** SEM images of partially degraded PDC multiblock copolymer with 67 wt % PCL content after 24h, 72 and 100 h exposure to phosphate buffered (pH 7.2) Pseudomonas lipase solution (200 $\mu g \cdot ml^{-1}$) at 37 °C.

In combination with the observation that the degradation solution began to turn turbid after 24 h and the overall weight loss of these samples after 100 h immersion was higher than the initial PCL content, it could be concluded that surface erosion of microparticles is a major process of the selective enzymatic degradation of PDC multiblock copolymers. In case of a hydrolytic degradation no changes of the surface occurred [7]. This finding might be explained by the co-existence of different domains inside the PDC polymers with different stability against enzymatic attack of lipase. Firstly, the amorphous PCL domains and secondly the PCL crystallites are degraded [7], while amorphous and crystalline PPDO domains segments remain unaffected. Finally, porous PDC microparticles get released and immersed in the buffer solution.

## CONCLUSIONS

Most of the PDC degradation products found in the periimplantary tissues had a diameter of 1 – 3 micron. This in good agreement with data obtained in *in vitro* studies, in which the enzymatically degradation of PDC was analyzed. The surface roughness increased because of enhanced erosion and microparticles were visible on the polymer surface. The study showed that the *in vivo* degradation process of PDC is characterized by an release of microparticles. The in vivo experiments resulted, that PCL is degraded faster than PPDO. Therefore it is speculated that the formed particles primarily consisted of PPDO. Such the strong microparticular degradation of the polyetherester is expected to support soft tissue integration better than the degradation behavior of the pure polyesters. This will be studied in future investigations.

## ACKNOWLEDGMENTS

Authors are grateful to the German Federal Ministry of Education and Research for providing financial support within the BioFuture Award No. 0311867.

# REFERENCES

1. D. Hofmann, M. Entrialgo, K. Kratz, A. Lendlein, „Knowledge-based approach towards hydrolytic degradation of polymer based biomaterials", *Adv. Mat.* DOI:10.1002/adma.200802213, published online May 4 2009.
2. A. Lendlein, S. Kelch, *Angewandte Chemie International Edition* **41**, 2034 (2002).
3. M. Behl, A. Lendlein, *Soft Matter* **3**, 58 (2007).
4. A. Lendlein, R. Langer, *Science* **296**, 1673 (2002).
5. K. Kratz, U. Voigt, W. Wagermaier, A. Lendlein, in Advances in Material Design for Regenerative Medicine, Drug Delivery, and Targeting/Imaging, edited by V. P. Shastri, A. Lendlein, L-S. Liu, A. Mikos, S. Mitragotri 1140-HH03-01, Mater. Res. Soc. Symp. Proc. Volume 1140E, Warrendale, PA (2009).
6. R. Mohr, K. Kratz, T. Weigel, M. Lucka-Gabor, M. Moneke, A. Lendlein, *Proceeding of the National Academy of Science of the United States of America* **103**(10), 3540-3545 (2006).
7. A. Kulkarni, J. Reiche, J. Hartmann, K. Kratz, A. Lendlein, *European Journal of Pharmaceutics and Biopharmaceutics* **68**(1),46-56 (2008).
8. J. Reiche, A. Kulkarni, K. Kratz, A. Lendlein, *Thin Solid Films* **516**, 8821-8828 (2008).
9. Bakker D, van Blitterswijk CA, Hesseling SC, Daems ThW, Grote JJ. *J. Biomed Mater Res* **24**: 277 (1990).
10. Zajaczkowski MB, Cukierman E, Galbraith CG, Yamada KM. *Tissue Eng* 2003; **9**: 525 (2003)

Mater. Res. Soc. Symp. Proc. Vol. 1190 © 2009 Materials Research Society 1190-NN10-02

# Surface Coatings Based on Polysilsesquioxanes: Grafting-From Approach Starting From Organic Polymers

Daniel Kessler and Patrick Theato
Institute of Organic Chemistry, Johannes Gutenberg University Mainz, Duesbergweg 10-14, 55099 Mainz

## ABSTRACT

Poly(methylsilsesquioxane) (PMSSQ) based hybrid materials are promising candidates to produce substrate-independent stable and adherent surface coatings. Usually these materials are synthesized by controlled radical polymerization from inorganic precursors. The presented synthetic pathway in here demonstrates how to graft PMSSQ networks from an endgroup-functionalized organic polymer and thus enlarges the range of accessible inorganic/organic hybrid coating materials.

## INTRODUCTION

The quality of adhesion between solids, e.g. between a film and its substrate, depends to a large extent on the microstructure of the interface layer that is being formed [1]. Usually four types of interface layers can be formed during the coating process of polymers onto different substrates: 1. mechanically interlocked interfaces; 2. chemical bonding interface layers; 3. electric double layers; 4. diffusion interface layers [2,3].

Many different approaches have been explored recently to coat various substrates with polymers for various applications: The technique of grafting polymers from or onto surfaces [4-6] relies on chemical bonding interface layers and therefore depends on corresponding functional groups on the surface. Sol-Gel based coatings require free hydroxyl groups on the surface to form chemical bonds. For example on glass, silicon or metals hydroxyl groups are usually present [7], while on other substrates an additional plasma procedure may be necessary [8]. Polymer coatings on plastic substrates adhere due to a diffusion interface layer, if the polymers are miscible [9], requiring fine tuning between film and surface material. Electric double layer interfaces were formed in polyelectrolyte coatings [10]. Cross-linkable random copolymers as coating materials [11] exhibit a mechanically interlocked interface on rough surfaces. After cross-linking the former substrate topography is filled with rigid and stable polymer networks.

The disadvantage of almost all reported procedures is their limitation towards the natured substrate to be coated, which is due to the utilization of only one interface layer phenomenon to guarantee adhesion.

Recently, we introduced inorganic/organic hybrid polymers, consisting of poly(methylsilsesquioxane) (PMSSQ) and radically polymerized organic polymers [12,13]. After curing, stable and adherent films on various substrates have been achieved. On hydroxylated surfaces chemical bonding interfaces causes adherence, on polymeric substrates diffusion interfaces improve adherence and due to cross-linking of the PMSSQ moieties rigid polymer networks at the interface create mechanical interlocking.

Usually those hybrid polymers were synthesized in a controlled radical polymerization of the organic monomer from a PMSSQ precursor (ATRP in ref.[12] and RAFT in ref. [13]), which limits this class of materials to (meth)acrylate or styrene based polymers. Within these studies we want to enlarge the range of accessible PMSSQ hybrid coating materials by grafting PMSSQ from the pre-formed organic polymer.

## EXPERIMENT

**Materials.** All chemicals and solvents were commercially available (Acros Chemicals, ABCR) and used as received unless otherwise stated. THF was distilled from sodium/benzophenone under nitrogen.

**Instrumentation.** $^1$H-NMR spectra were recorded on a Bruker 300 MHz FT-NMR spectrometer, $^{29}$Si CPMAS NMR spectra were measured on a Bruker DSX 400 MHz FT-NMR spectrometer (Rotation: 5000 Hz, T = RT, 4 mm rotor). Chemical shifts ($\delta$) were given in ppm relative to TMS. Gel permeation chromatography (GPC) was used to determine molecular weights and molecular weight distributions, $M_w/M_n$, of polymer samples. (THF used as solvent, polymer concentration: 2 mg/mL, column setup: MZ-Gel-SDplus 102 Å2, 104 Å2 and 106 Å2, used detectors: refractive index, UV and light scattering). Thermo gravimetrical analysis was performed using a Perkin Elmer Pyris 6 TGA in nitrogen (10 mg pure polymer in aluminum pan). Atomic force microscopy (AFM) measurements were performed using a Veeco Dimension 3100 in tapping mode. IR spectra were recorded using a Nicolet 5 DXC FT-IR-spectrometer on ATR crystal. Advancing and receding CA of water were measured using a Dataphysics Contact Angle System OCA 20 and fitted by SCA 20 software. Given CA are average values of 10 individual measurements with an accuracy of 3°. All chemical reactions were performed in Argon atmosphere.

**Allylphenyl-terminated Polycarbonate (1)** was synthesized as explained in [14]. Yield: 44.2 g. $^1$H-NMR (CDCl$_3$) $\delta$ (ppm): 7.25 (d, 60H); 7.17 (d, 60H); 5.93 (m, 2H); 5.12 (m, 4H); 3.43 (d, 4H); 1.68 (s, 77H). $M_n$ = 4912 g/mol; PDI = 1.38. $T_g$ = 180 °C.

**Trichlorosilyl-propylphenyl-terminated Polycarbonate (2).** 15 g of (1) was dissolved in 100 mL THF and 15 mg platinum on charcoal and 10 mL silicochloroform were added. The solution was refluxed at 80 °C for 12 h. Afterwards the solution was filtered over celite and the solvent was removed under reduced pressure. Yield: 14.7 g. $^1$H-NMR (CDCl$_3$) $\delta$ (ppm): 7.26 (m, 60H); 7.17 (d, 60H); 2.77 (t, 4H); 1.67 (s, 81H); 0.92 (m, 4H).

**Poly(methylsilsesquioxane)-functionalized Polycarbonate (3).** 4 g of (2) and 4.42 g methyl trimethoxysilane were dissolved in 250 mL THF and 2.93 mL water and 0.24 mL 2N HCl were added. The solution was stirred at room temperature for 8 h. 300 mL chloroform was added, the solution was washed three times with water, dried over MgSO$_4$ and the solvents were removed. The polymer was dried in high vacuum. Yield: 5.8 g. $^1$H-NMR (CDCl$_3$) $\delta$ (ppm): 7.27-7.09 (m); 5.02 (br); 2.78 (t, 4H)); 1.67 (s, 81H); 0.91 (m, 4H); 0.16 (br, 29H). $^{29}$Si CPMAS NMR $\delta$ (ppm): -48.61 (T1, 4%); -57.20 (T2, 58%); -66.08 (T3, 38%). $M_n$ = 47200 g/mol; PDI = 2.15.

**Di-(trichlorosilyl)-functionalized Polyethyleneglycol (4).** 5 g polyethyleneglycol (2000 g/mol, water removal by azeotrope distillation with toluene) was dissolved in 100 mL THF and 1 g isocyanatopropyltriethoxysilane and 0.5 mL triethylamine were added. The mixture was stirred at room temperature for 12 h. After removing the solvent at low pressure, the polymer was dried in high vacuum. Yield: 4.55 g. [1]H-NMR (CDCl$_3$) δ (ppm): 6.30 (s, 2H); 4.12 (m, 4H); 3.79 (quad., 12H); 3.61 (br, 240H); 3.11 (quad., 4H); 1.61 (m, 4H); 1.18 (t, 18H); 0.60 (t, 4H).

**Poly(methylsilsesquioxane)-functionalized Polyethylenglycol (5).** 4 g of (4) and 4.42 g methyl trimethoxysilane were dissolved in 250 mL THF and 2.93 mL water and 0.24 mL 2N HCl were added. The solution was stirred at room temperature for 8 h. 300 mL chloroform was added, the solution was washed three times with water, dried over MgSO$_4$ and the solvents were removed. The polymer was dried in high vacuum. Yield: 5.2 g. [1]H-NMR (CDCl$_3$) δ (ppm): 6.34 (s); 4.12 (m); 3.61 (br, 240H); 3.11 (quad.); 1.61 (m); 0.61 (t); 0.15 (br, 186H). [29]Si CPMAS NMR δ (ppm): -47.98 (T1, 6%); -58.01 (T2, 54%); -65.55 (T3, 40%). M$_n$ = 38800 g/mol; PDI = 2.44.

**Surface coating.** The hybrid polymer solution was spin-coated onto clean substrates (15 s, 4000 rpm, 10 wt% solution in THF). To induce the secondary cross-linking of the inorganic block, the samples were annealed at 130 °C for 2 h and afterwards washed with THF for 30 minutes to remove any non-bonded material.

## DISCUSSION

PMSSQ-based hybrid polymers obtained in an ATRP or RAFT process show an excellent adherence on various substrates (e.g. Si, glass, copper, steel, gold, PMMA and PDMS) [12,13]. This method gives convenient access to defined surface properties on different substrates (e.g. to coat gold with PMMA using a PMSSQ-PMMA hybrid polymer). Stable and adherent polymeric coatings of non-radically formed polymers like polycarbonate (PC) (produced via polycondensation) or polyethyleneglycole (PEG) (produced via anionic polymerization) could not been produced by covalent attachment of those polymers to PMSSQ.

### Synthesis of PMSSQ-PC and PMSSQ-PEG hybrid materials

Our synthetic concept to attach pre-formed polymers to a PMSSQ network consists of a trichlorosilyl-endgroup functionalization of PC or PEG and further co-condensation with methyl trimethoxysilane (MTMS) to yield a PMSSQ network (synthetic scheme for PMSSQ-PC hybrid materials: figure 1).

To synthesize endgroup-functionalized PC (2), we quenched the polycondensation reaction of bisphenol A and diphosgene with o-allylphenol and the successful introduction of the allyl-endgroups was confirmed by [1]H NMR spectroscopy (figure 1, upper spectrum). Hydrosilylation with silicochloroform and further co-condensation with MTMS yielded the desired PMSSQ-PC hybrid polymer (3). [1]H NMR spectroscopy of (3) shows the additional Si-CH$_3$ signal at 0.16 ppm, whereas the allyl signals vanished (figure 1, lower spectrum). The molecular weight increased from 5 kg/mol to 47 kg/mol and showed a monomodal distribution, indicating that PC segments were incorporated in a PMSSQ network. The ability of this hybrid material to undergo further cross-linking was checked by [29]Si CPMAS NMR spectroscopy, 62% of all silicon atoms still carry two or one free OH group (T1 and T2 branches). Cross-linking occured, similar to other PMSSQ systems, between 100 °C and 130 °C and took less than 70 min at 130 °C (measured by TGA). The ratio between silicon moieties and organic repeating units, calculated by NMR integration, was 50:50. Taking molecular weights into account the weight ratio between inorganic and organic part was calculated as 27% : 73 %.

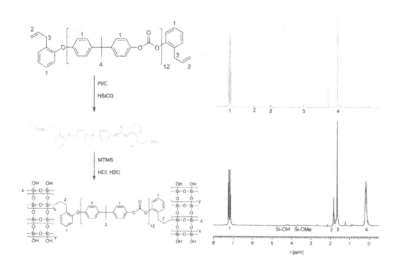

**Figure 1.** Synthetic pathway toward PMSSQ-PC hybrid polymers, including $^1$H NMR spectra.

A PMSSQ-PEG hybrid polymer was synthesized by endgroup-functionalization of PEG-diol (2 kg/mol) with isocyanatopropyltriethoxysilane, yielding - similar to the functionalization of PC - a di-(trichlorosilyl)-functionalized PEG (4). Further co-condensation with MTMS led to the PMSSQ-PEG hybrid polymer (5). GPC showed a monomodal molecular weight distribution with $M_n$ = 39 kg/mol. $^{29}$Si CPMAS NMR spectroscopy showed 60% T1 and T2 branches, cross-linking occurred under similar conditions as explained above. The ratio between silicon moieties to PEG repeating units was calculated to 50:50. Weight ratio between inorganic and organic block was calculated as 72% : 28%.

PMSSQ-PC (3) as well as PMSSQ-PEG (5) showed the desired features to produce stable and adherent surface coatings after spin-coating and annealing.

## Preparation of PMSSQ-PC and PMSSQ-PEG coatings on various substrates

10 wt% solutions of PMSSQ-PC (3) and PMSSQ-PEG (5) in THF were spin-coated at 4000 rpm onto Si, glass, copper, steel, gold, PMMA, PC and PDMS substrates, respectively. The coated substrates were cured at 130 °C for 2 h to thermally induce cross-linking of the PMSSQ part.

To characterize the surface topography and film thickness, AFM measurements were performed on coatings on silicon wafers. The PMSSQ-PC coating showed an image RMS value of 0.523 nm , indicating a smooth coating. The film thickness was 420 nm. The PMSSQ-PEG coating resulted in smooth surfaces (image RMS: 0.492 nm), film thickness was 375 nm.

The successful surface modification was tested by measuring advancing ($\Theta_a$) and receding ($\Theta_r$) contact angles on all different substrates. After coating with PMSSQ-PC the advancing contact angle was 78 ° ± 5 °, on all coated materials (see table I). To demonstrate the long-term stability of the coating, the wafers were kept in boiling water for 4 h and the contact angle was checked every hour; the surface properties were not affected, indicating a substrate-independent permanent surface modification (table I). The adhesion and stability of the coatings on the underlying material was tested using the ISO tape test [15] directly after annealing and in a similar long-term stability experiment as explained above. Classification 0 indicates a perfect adherence (no detachment), classification 5 indicates more than 50 % detachment of the coating. The observed stabilities on all substrates are summarized in table I, even after 4 h in boiling water the tape test result did not drop below classification 1, indicating the desired high stability on a wide range of materials.

**Table I.** Contact angles and tape test results of PMSSQ-PC coatings on various substrates.

| | $\Theta_a/\Theta_r$ [°] | | | | | tape test results | | | | |
|---|---|---|---|---|---|---|---|---|---|---|
| | 0 h | 1 h | 2 h | 3 h | 4 h | 0 h | 1 h | 2 h | 3 h | 4 h |
| Si | 79/69 | 78/69 | 79/70 | 77/70 | 78/70 | C | 0 | 0 | 0 | 0 |
| Glass | 77/69 | 78/67 | 78/65 | 77/69 | 77/70 | 0 | 0 | 0 | 0 | 0 |
| Copper | 82/71 | 81/71 | 82/74 | 82/70 | 81/70 | 0 | 0 | 0 | 1 | 1 |
| Steel | 83/70 | 82/70 | 80/70 | 81/70 | 80/71 | 0 | 0 | 0 | 1 | 1 |
| Gold | 80/74 | 79/72 | 78/70 | 79/71 | 77/69 | 0 | 0 | 1 | 1 | 1 |
| PMMA | 74/68 | 75/68 | 74/70 | 73/70 | 74/69 | 0 | 0 | 0 | 0 | 1 |
| PC | 73/69 | 74/69 | 75/73 | 73/68 | 74/68 | 0 | 0 | 0 | 0 | 1 |
| PDMS | 73/70 | 75/68 | 77/70 | 77/69 | 77/70 | 0 | 0 | 1 | 1 | 1 |

Similar experiments were performed to determine the successful surface coating using PMMSQ-PEG hybrid materials. Advancing contact angles on all tested underlying materials were 39 ° ± 5 °, which did not change significantly after 4 h in boiling water independent from the substrate. Tape test results also support the high long-term stability of the PMSSQ-PEG coating (for detailed information see table II).

**Table II.** Contact angles and tape test results of PMSSQ-PEG coatings on various substrates.

| | $\Theta_a/\Theta_r$ [°] | | | | | tape test results | | | | |
|---|---|---|---|---|---|---|---|---|---|---|
| | 0 h | 1 h | 2 h | 3 h | 4 h | 0 h | 1 h | 2 h | 3 h | 4 h |
| Si | 41/32 | 40/35 | 42/37 | 40/36 | 40/35 | 0 | 0 | 0 | 0 | 0 |
| Glass | 39/34 | 40/34 | 38/34 | 38/32 | 38/33 | 0 | 0 | 0 | 0 | 0 |
| Copper | 34/29 | 35/29 | 36/31 | 35/30 | 35/30 | 0 | 0 | 1 | 1 | 1 |
| Steel | 34/27 | 34/28 | 35/28 | 34/27 | 34/27 | 0 | 0 | 0 | 1 | 1 |
| Gold | 35/29 | 35/31 | 34/31 | 35/31 | 36/30 | 0 | 1 | 1 | 1 | 1 |
| PMMA | 42/35 | 41/34 | 40/36 | 41/36 | 39/36 | 0 | 1 | 1 | 1 | 1 |
| PC | 41/36 | 43/36 | 40/35 | 39/33 | 40/34 | 0 | 0 | 0 | 1 | 1 |
| PDMS | 40/31 | 39/34 | 38/32 | 39/33 | 38/33 | 0 | 0 | 0 | 0 | 1 |

Both coating materials, PMSSQ-PC as well as PMSSQ-PEG, showed comparable abilities to produce substrate-independent adherent surface coatings with high long-term stabilities.

## CONCLUSIONS

PMSSQ-activated hybrid materials successfully combine several interlayer phenomena in one coating material and thus form stable and adherent coatings on a wide range of substrates. The presented synthetic scheme allows preparing PMSSQ-based inorganic/organic hybrid materials starting from pre-formed organic polymers, using first a trichlorosilyl-functionalization of both polymer endgroups and a polycondensation with MTMS yielding the desired hybrid PMSSQ network. PMSSQ-PC and PMSSQ-PEG could be used to produce surface coatings on a wide range of substrates. Successful surface modification and high long-term stability of the coatings could be demonstrated by contact angle measurements and ISO tape test.

## ACKNOWLEDGMENTS

D.K. gratefully acknowledges financial support from the FCI, POLYMAT (Graduate School of Excellence "Polymers in Advanced Materials") and the IRTG 1404 ("Self-Organized Materials for Optoelectronic Applications").

## REFERENCES

1.  H. K. Pulker, A. J. Perry, Surface Technology **14**, 25 (1981).
2.  W. Possart (Ed.), "Adhesion. Current Research and Applications" 2005, Wiley-VCH (Weinheim).
3.  D. E. Packham (Ed.), "Handbook of Adhesion" 2$^{nd}$ Ed. 2005; John Wiley & Sons (New York).
4.  P. Mansky, Y. Liu, E. Huang, T. P. Russell, C. J. Hawker, Science **275**, 1458 (1997).
5.  R. C. Advincula, W. J. Brittain, K. C. Caster, J. Rühe, (Eds.) "Polymer Brushes" 2004, Wiley-VCH (Weinheim).
6.  L. Gao, T. J. McCarthy, J. Am. Chem. Soc. **128** (28), 9052 (2006).
7.  J. den Toonder, J. Malzbender, G. de With, R. Balkenende, J. Mater. Res. **17** (1), 224 (2002).
8.  C. M. Chan, G. Z. Cao, H. Fong, M. Sarikaya, T. Robinson, L. Nelson, J. Mater. Res. **15** (1), 148 (2000).
9.  A. R. Marrion (Ed.), "The Chemistry and Physics of Coatings" 2004, RSC (Cambridge).
10. N. Benkirane-Jessel, P. Lavalle, V. Ball, J. Ogier, B. Senger, C. Picart, P. Schaaf, J.-C. Voegel, G. Decher, Macromol. Eng. **2**, 1249 (2007).
11. D. Y. Ryu, K. Shin, E. Drockenmuller, C. J. Hawker, T. P. Russell, Science **308**, 236 (2005).
12. D. Kessler, C. Teutsch, P. Theato, Macromol. Chem. Phys. **209** (14), 1437 (2008).
13. D. Kessler, P. Theato, Macromolecules **41** (14), 5237 (2008).
14. S. H. Kim, H. G. Woo, S. H. Kim, H. G. Kang, W. G. Kim, Macromolecules 32, 6363 (1999)
15. Paints and Varnishes – Cross-cut test (ISO 2409:2007)

Mater. Res. Soc. Symp. Proc. Vol. 1190 © 2009 Materials Research Society          1190-NN11-23

# *In vitro* study of mouse fibroblast tumor cells with TNF coated and Alexa488 marked silica nanoparticles with an endoscopic device for real time cancer visualization

Marion Herz[1], Andreas Rank[1], Günter E. M.Tovar[1,2], Thomas Hirth[1,2], Dominik Kaltenbacher[3], Jan Stallkamp[3], Achim Weber[1,2]

[1]Fraunhofer-Institute for Interfacial Engineering and Biotechnology,
[2]Institute of Interfacial Engineering, University of Stuttgart,
[3]Fraunhofer Institute for Manufacturing Engineering and Automation
Nobelstr. 12, 70569 Stuttgart, Germany

## ABSTRACT

Tumor resection done by minimally invasive procedure owns the challenge of a fast and reliable differentiation between healthy and tumorous tissue. We aim at investigating and developing a method for an intraoperative visualization of tumor cells with functionalized nanoparticles. The goal is to use this technique for the intraoperative use. Our so-called biohybrid systems consist of nanoparticles that are produced by Stöber synthesis and coupled with bio active proteins. Such biomimetic nanostructures are capable of imitating the effects of membrane-bound cytokines, which bind to tumor cells for labeling them. A flexible and modular test environment has been developed to evaluate the spraying properties of the particles and to study tissue probes. It enables a fast investigation of different particle configurations and spraying parameters like pressure, spray volume, nozzle geometry, etc.

## INTRODUCTION

During the course of minimally invasive tumor resection, it is often difficult for the surgeon to clearly distinguish between healthy and tumorous tissue on the endoscopic image. Furthermore, the surgeon does not perceive the haptic and tactile feedback needed to evaluate the tissue by means of palpatation. The use of navigation systems to localize the tumor usually fails due to the presence, quality, accuracy or the informational content of the image data. Therefore a tumor sample is removed during surgery and given to the pathologist. This procedure can take up to half an hour and requires that the operation is "put on hold" for this period of time. However, a final laboratory examination can take significantly longer. To minimize the risk, the surgeon supplements the expertise of the pathologist and a mutual assessment of the tissue status is usually done. This decision is extremely risky, for example, if the partial resection of a tissue or an entire organ depends on it. Quality control of the resection margins is so far not possible by any means. Thus, the necessity of the resection of large volumes remains for reasons of safety. The C-VIS idea concerns with an alternative method which in a short period of time can directly help to visualize the tumor tissue in human body.

## EXPERIMENTAL

### Making tumor tissue visible with nanoparticles

The C-VIS method is based on the property of modified nanoparticles attaching to tumor cells, but not to healthy cells. Nanoparticle hybrid systems are synthesized consisting of synthetic nanoparticles and biologically active proteins. For example these biomimetic constructs are able to mimic the specific properties of membrane bound cytokine (e.g. Tumor Necrosis Factor). The proteins are genetically modified to induct specific functionalities as an anchor site at well defined positions. For that reason nanoparticles are produced by means of colloid chemistry and are provided with surface functions that are complementary to those of the introduced protein reactivities. The protein constructs are then conjugated with the nanoparticles in a chemically mild and thus protective reaction [1].

### Unambiguous marking during the operation

During surgery, these particles should be sprayed through the endoscope onto the tissue surface or the resection margins. The particles which have previously been treated with fluorescent markers are then stimulated by a light source at the appropriate wave length and thus made visible in contrast to the healthy tissue. On the video endoscope image the surgeon can see the stimulated and passive surface segments *in vivo*. With this information, he can treat the tumor and the resection in a targeted manner. Figure 1 illustrates the idea of C-VIS.

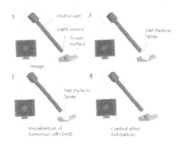

Figure 1: The progress of an operation:
**1.** Positioning of the endoscope, **2.** Spraying of the nano-particles: the marked TNF-functionalized nanoparticles bind to cells, **3.** Intraoperative tumor visualization, **4.** Control Adapted from [2].

### Synthesis of nanoparticles and deposition onto substrates

Spherical core-shell nanoparticles are generated by the method of Stöber by alkaline hydrolysis and condensation of tetraethoxysilane. This method turns out to be relatively simple whereas the silica particles are chemically inert. Surface modification reactions by silanization provide a high flexibility regarding the type of functional groups, their density and when needed their spacer length. For example a conversion with functional trialkoxysilanes leads to particles with covalently bound amino groups at their surface [3]. With ordinary co-condensations

126

additional functions like fluorescent markers, radio nuclides, etc. can covalently be integrated into the nanoparticles [1]. Furthermore nanostructured particles with tailor-made affinity surfaces were generated by bioconjugation reactions [1,3,4,5]. An increasing miniaturization into the nanometer range of the affine and spherical carrier systems enhances the specific surface over again and thereby magnifies the sensitivity.

Glass slides are used as substrates for the arrangements of the nanoparticles which in a first experiment are sprayed onto their surface. The glass slides are cleaned with a 2% Hellmanex® solution for 45 minutes at 40°C. Hydroxylation of the surface was proceeded immediately by dip coating into a 3:1 (v/v) $NH_3$ solution (25 wt. %) and $H_2O_2$ (30 wt. %) for 40 minutes at 70°C. After every step in the coating process several times rinsing in ultrapure water is required. For being able to discriminate between two different surface properties the already cleaned and hydroxylated substrates are coated with positively and negatively charged polyelectrolytes (poly(diallyldimethyl) ammonium chloride, PDADMAC ($MW_{Monomer}$ = 1,32 Da) and the sodium salt of poly(styrene sulfonate), SPS($MW_{Monomer}$ = 206 Da)) [3].

A multilayer of PDADMAC (0.02 M referring to the monomer) and SPS (0.01 M referring to the monomer) has been performed via *Layer-by-Layer* technique by dipping the substrates for 20 minutes into the particular poly-electrolytic solution whereas unspecific electrostatic interactions of the differently charged polyelectrolyte molecules are used to generate a desired surface charge. There is no charge balancing in this process because of the hydrated and flexible polymer chains that don't apply parallel to the surface but form polymer loops which extend into the liquid phase and create a plenty of free charge carriers. As nanoparticles are charged as well they can find stable binding sites at the substrates' surface without losing their biological activity [7,8]. These substrates are then irradiated with deep UV light through a grid mask and the poly-electrolytic layer is thus partially removed and chemically modified. The nanoparticles are able to self-organize between the modified surface and the untreated area whereby a positive image of the mask is received [9].

*Endoscopic instrument*
The design of this novel endoscope shows high flexibility and modularity. In Figure 2 the experimental set up is shown. The nanoparticle suspension with a solid content of 10 mg/mL (m/v) in acetic buffering solution at pH 4.7 is pressurized through the atomizing nozzle according to the principle of a syringe. The best results were purchased with a pressure of 4 bar and a speed rate for spraying of 400 Hz. A high frequency plug valve and the precise adjustable compressed air supply ensure accurate quantitative and prompt dosage of the nanoparticles. A standardized interface between valve and atomizing nozzle plus fixing of the particle reservoir by a quick clamp device are additional features of the set up which enables simple examinations of variable nozzle types and geometries. With the quick clamp device it is possible to perform fast and flexible changes of the particle reservoir and thus spray with various suspensions. All mentioned components have been chosen in respect of later miniaturization and integration into an endoscopic device.

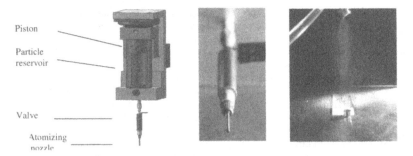

Piston

Particle
reservoir

Valve

Atomizing
nozzle

Figure 2: **Left**: Design, **Middle**: Image of the endoscopic unit. **Right**: Picture of an ongonig spraying experiment to a substrate [2].

### RESULTS AND DISCUSSION

*Generation of micro-structured surfaces*

Silica nanoparticle suspension is extensively sprayed onto glass substrates which are already cleaned and coated with polyelectrolytes. Before spraying the desired assembly of the nanoparticles is achieved by structuring the polyelectrolytic coating photolitographically with deep UV light [9]. This leads to areas where the polyelectrolytes are removed and chemically modified. Electrostatic forces between the surface of the substrate and the charged nanoparticles allow their stable binding only at the untreated area which is illustrated in Figure 3. Unbound particles are washed off and the pattern of the grid is clearly presented.

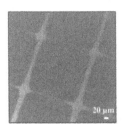  

Figure 3: SEM images of a pretreated sample after spraying with nanoparticles and subsequent rinsing steps. It is clear that the nanoparticles only attach to the intended locations. From left to right, magnifications of 500; 2,500 and 25,000 are shown.

*Induction of Apoptosis in MF-R1-Fas and MF-R2-Fas Cells*

Membrane bound TNF (mTNF) and soluble TNF (sTNF) bind to two different cell membrane receptors 1 and 2. mTNF strongly activates both receptors, whereas sTNF is able to only stimulate TNFR1 but not TNFR2. TNF-modified and Alexa-Fuor-488 labeled nanoparticles with a diameter of 100 nm were incubated with mouse fibroblasts (MF) to test the specific bioactivity of the labeled nanoparticles. Figure 4 shows the development of cell death in cells transfected with receptor 1 and receptor 2.

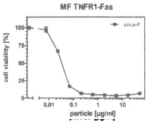 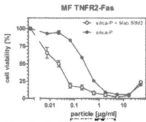

Figure 4: Receptor selectivity of TNF-coated silica nanoparticles. Modified nanoparticles with a recombinant TNF protein bind to and strongly activate both receptors, as a result mimicking the character of mTNF. **Left**, Apoptosis occurs after treatment with nanoparticles at concentrations of 0.1 µg/ml. **Right**, Apoptosis can be observed, too after treatment with nanoparticles at concentrations of 0.5 µg/ml. In presence of red fluorescent antibody 80M2 specific for TNFR2 apoptosis occurs even faster at a lower particle concentration. [Tests were performed by Sylvia Messerschmidt, Cooperation with Prof. Scheurich, Institute for Cell Biology and Immunology IZI, University of Stuttgart]

Nanoparticles modified with TNF are also able to specifically bind to the MF TNFR2-cells and induce apoptosis whereas this receptor is fully resistant to soluble TNF activity. The binding of Alexa-Fluor-488 stained nanoparticles to mouse fibroblasts treated with Alexa Fluor 546-labeled red fluorescent antibody 80M2 was detected with confocal fluorescence microscopy. After 15 minutes reaction time the cells show specific receptor aggregation.

 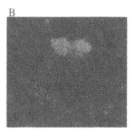 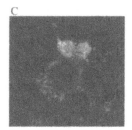

Figure 5: Confocal fluorescence microscope images of mouse fibroblasts with TNF-marked nanoparticles. The cells were observed after staining and fixing them onto glass slides for several days. **A**, Mouse fibroblasts treated with Alexa Fluor 546 labeled red fluorescent antibody 80M2. **B**, Alexa Fluor 488 stained nanoparticles with a diameter of 100 nm, coated with TNF. **C**, Overlay of both pictures, yellow clusters show binding of the nanoparticles to the cells' surface.

## CONCLUSIONS

The C-VIS project is an alternative or complementary solution to computer-assisted diagnosis (CAD) which presently focuses on approaches using technical analysis software. After an evaluation of the spraying process subsequently with tissue, preclinical studies are still outstanding. These studies will examine the procedure for its suitability in the actual practice of medicine. If this meets with success, the next step will focus on developing different nanoparticles to detect specific tumors, investigating solutions for automatic image evaluation and pursuing the options for precise spatial localization of the tumor. In the future C-VIS could be integrated into an automated resection procedure such as one that uses a robot system.

## ACKNOWLEDGMENTS

The authors thank Andreas Ritter and Sylvia Messerschmidt (Institute for Cell Biology and Immunology, IZI) for lab work, Monika Riedl (Fraunhofer IGB) for SEM images, Margarete Witkowsky (IZI) for confocal microscopy images and the Fraunhofer Gesellschaft and Johnson & Johnson for funding.

## REFERENCES

1. Bryde S., Grunwald I., Hammer A. et al, *Tumor Necrosis Factor (TNF)-Functionalized Nanostructured Particles for the Stimulation of Membrane TNF-Specific Cell Responses.* Bioconjugate Chemistry, 2005. **16**: p. 1459-1467.
2. A. Weber, M. Herz, T. Hirth, et al. *C-VIS: Interoperative Tumorerkennung mit Hilfe von Nanopartikeln,* Endoskopie heute, 2009, **22**: p. 36-39.
3. Tovar GEM, Scheurich P., *Nanotechnologische Werkzeuge für die Immunologie.* Biowolrd, 2002.**1**: p. 6-7.
4. Stöber W., Fink A., Bohn E., *Controlled growth of monodisperse silica spheres in the micron size range.* Journal of Colloid and Interface Science, 1968. **26**: p. 62.
5. Schiestel T., Brunner H., Tovar GEM., *Controlled surface functionalizations of silica nanospheres by covalent conjugation reactions and preparation of high density streptavidin nanoparticles.* Journal of Nanoscience and Nanotechnology, 2004. **4**: p. 504-511.
6. Borchers K., Weber A., Brunner H. et al, *Microstructured layers of spherical biofunctional core-shell nanoparticles provide enlarged reactive surfaces for protein microarrays.* Analytical and Bioanalytical Chemistry, 2005. **383**: p. 738-746
7. Borchers K., Weber A., Hiller E. et al, *Nanoparticle-based diagnostic 3D-protein-biochip for candida albicans.* PMSE Preprints, 2006. **95**: p. 1016-1017.
8. Decher G., *Fuzzy nanoassemblies: toward layer polymeric multicomposites.* Science, 1997. **277**: p. 1232-1237.
9. Tovar GEM., Weber A., *Bio-Microarrays based on Functional Nanoparticles.* Dekker Encyclopedia of Nanoscience and Nanotechnology. New York: Marcel Dekker, Inc. 2004. p. 277-286.

Mater. Res. Soc. Symp. Proc. Vol. 1190 © 2009 Materials Research Society    1190-NN11-34

# Amorphous Polymer Networks Combining Three Functionalities - Shape-Memory, Biodegradability, and Drug Release

Christian Wischke[1,2], Axel T. Neffe[1], Susi Steuer[3], Andreas Lendlein[1,2]

[1] Center for Biomaterial Development, Institute for Polymer Research, GKSS Research Center Geesthacht GmbH, Kantstrasse 55, 14513 Teltow, Germany
[2] Berlin-Brandenburg Center for Regenerative Therapies, Berlin, Germany
[3] present address: Intervet Innovation GmbH, 55270 Schwabenheim, Germany

## ABSTRACT

Shape-memory polymers are of high scientific and technological interest in the biomedical field, e.g., as matrix for self-anchoring implantable devices. In this study, two different star-shaped copolyester tetroles, semi-crystalline oligo[(ε-caprolactone)-co-glycolide]tetrol (oCG) and amorphous oligo[(rac-lactide)-co-glycolide]tetrol (oLG), were synthesized and subsequently crosslinked by a low molecular weight diisocyanate resulting in copolyester urethane networks (N-CG, N-LG). Both networks could be loaded with model drugs and a diffusion controlled release of the drugs was observed without any effect on the mass loss as measure of hydrolytic degradation. However, the N-CG network's capability of shape programming was disturbed as the crystallinity of the precursors got lost in the complex three dimensional architecture after crosslinking. By contrast, amorphous N-LG network showed an excellent shape-memory capability with a switching temperature around 36 °C corresponding to their glass transition temperature. This led to triple-functional materials combining biodegradability, shape-memory, and controlled drug release.

## INTRODUCTION

Shape-memory polymers (SMP) [1] belong to the class of 'actively moving' polymers and possess the capability to recover from a temporarily fixed deformation (temporary shape) to their original shape. Different applications have been suggested for such SMP, either degradable or non-degradable materials, in the biomedical field including stents [2], as self-tightening sutures [3], intelligent electrodes [4], or thrombectomy devices [5]. In many cases, degradable SMP will be advantageous in order to avoid a second surgery for explantation. In order to initiate shape recovery, direct heating, indirect heating in a magnetic field [6,7], and light [8] have been described so far as suitable stimuli. The ability of a polymer to respond to such stimuli bases on the combination of the polymer's molecular architecture and morphology combined with a thermomechanical process for creating the temporary shape, called programming. On a molecular level, entropy-elastic polymer networks are required having either covalent or physical netpoints that determine the permanent shape. Besides the netpoints, a switching segment must be present, forming either semi-crystalline or amorphous domains related to a characteristic melting point $T_m$ or a glass transition $T_g$. Recently, semicrystalline polyester methacrylate networks [9] and amorphous copolyester urethane networks [10] were shown to beneficially combine controlled drug release, biodegradation, and shape memory capability and, therefore, led to multifunctional materials for biomedical applications. In this study, drug loading, drug release, and shape memory capability will be compared for polyester urethane networks build up from on the one hand semicrystalline and on the other hand amorphous tetrole precursors.

## EXPERIMENTAL DETAILS

Star-shaped oligo[(ε-caprolactone)-co-glycolide]tetroles (oCG) and oligo[(*rac*-lactide)-co-glycolide]tetroles (oLG) were synthesized by ring-opening polymerization with pentaerythrit (PE), a tetrole compound, as the starter molecule under nitrogen atmosphere at 130 °C for 8h (oCG) or 5d (oLG) [oCG: 235 mmol ε-caprolactone, 20.7 mmol diglycolide, 6 mmol PE, 0.6 mmol $(Bu)_2SnO$; oLG 2: 295 mmol rac-dilactide, 65 mmol diglycolide, 5 mmol PE, 0.4 mmol $(Bu)_2SnO)$]. Then, the reaction mixture was dissolved in a tenfold excess (v/w) of methylene chloride, precipitated by adding slowly a tenfold excess (v/v) of hexane, and washed with hexane three times. Subsequently, the precursors were dried at elevated temperature (oCG: 35 °C; oLG: 70 °C) and reduced pressure (1 mbar). For crosslinking, equimolar amounts of an isomeric mixture of 2,2,4 and 2,4,4-trimethylhexane-1,6-diisocyanate (TMDI) and the respective oligomeric precursor in methylene chloride were mixed in the presence of catalyst (N-CG: 0.8 mmol oCG, 1.6 mmol TMDI, 33.2 µl $Sn(Oct)_2$; N-LG: 8 mmol oLG, 16 mmol TMDI, 332 µl $Sn(Oct)_2$) and transferred into a nitrogen-flushed, vacuum-resistent container. After evaporation of methylene chloride, the container was closed and kept at 80 °C for 4 d in vacuo. For removal of non-crosslinked components, the networks were subjected to swelling in a 100fold excess of chloroform (v/w) for 24 h and dried at 35 (N-CG) or 60 °C (N-LG).

Multidetector gel permeation chromatographie (GPC) with universal calibration and 400 MHz $^1$H-NMR were used to determine the molecular weight and glycolide molar fraction $\chi_G$ of the oligomeric precursors. Differential scanning calorimetry (DCS) was performed on a DSC 7 equipped with a TAC 7/DX and CCA 2 (Perkin Elmer, Waltham, USA) using the data of the second heating run (from 0 to +100 °C with a heating rate of 10 K·min$^{-1}$) for further analysis. Tensile tests with n ≥ 3 repetitives were performed with dumbbell-shaped specimens at 37 °C in water (after a 4 h equilibration period) using a Zwick 2.5N1S machine. In an air atmosphere, tests were conducted at 70 °C (thermochamber) on a Zwick 1425 machine equipped with a Climatix 91259 thermochamber (Zwick GmbH & Co. KG, Ulm, Germany). The elastic modulus E, the elongation at breakage $\varepsilon_B$, and tensile strength $\sigma_{max}$ were obtained from stress-strain curves. Three different types of cyclic thermomechanic experiments were conducted; i) strain-controlled cycle in air (five cycles, ii) stress-controlled cycle in air (one cycle), and iii) stress-controlled cycle in water (one cycle).

Drug loading was performed by swelling of network samples in saturated solutions of enoxacin (EN) in $CH_3Cl$ or ethacridine lactate (EL) in 1:1 (m/m) $CHCl_3$ : 2-propanol with subsequent drying for solvent removal.

## RESULTS AND DISCUSSION
### Synthesis and Thermomechanical Properties of the Networks

Oligomeric, star-shaped tetroles were synthesized by ring opening polymerization of diglycolide and either ε-caprolactone or *rac*-dilactide with pentaerythrol as starter molecule. One major aim was to provide SMP with a thermal transition, i.e., $T_m$ for oCG and $T_g$ for oLG, that matches body temperature in order to allow shape recovery at physiological conditions. Therefore, oCG, being the N-CG precursor, was designed to have a low $M_n$ based on the assumption that this $M_n$ would reduce the $T_m$, which is 63 °C for poly($\varepsilon$-caprolactone), to 35 to 40 °C. In DSC measurements, oCG showed the predicted $T_m$ at about body temperature (see Table 1). The presence of two rather than one melting peak as observed for oCG may be explained either by the presence of crystallites of different thermodynamical stability or by artifacts from partial secondary crystallisation or recrystallisation, respectively, during melting in the DSC run. For oLG, a glycolide molar fraction $\chi_G = 18$ mol% was chosen in this study; whereas it is known generally that changing $\chi_G$ is a powerful tool to vary the degradation rate while affecting $T_g$ only to a minor extent.

Crosslinking the star-shaped precursors with diisocyanate to copolyester urethane networks resulted in the fixation of the oligomer arms in a complex network architecture. This led to a disturbed and therefore imperfect formation of crystallites with a reduced $T_m = 18$ °C due to limited chain mobility after crosslinking of the oligomers. More importantly, a strong reduction in crystallinity of N-CG from star-shaped oCG upon crosslinking can be assumed from the dramatic decrease in $\Delta H_m$ to less than 0.2% of its initial value (Table 1). For dry N-LG, an

**Table 1:** Characteristics of the polymer networks N-CG and N-LG as well as their precursors.

| | | N-CG | N-LG |
|---|---|---|---|
| **Oligomer properties** | Oligomer composition | oligo[($\varepsilon$-caprolactone)-co-glycolide]tetrole [oCG] | oligo[(*rac*-lactide)-co-glycolide]tetrole [oLG] |
| | $M_n$ ($^1$H-NMR) | 4,600 g·mol$^{-1}$ | 10,200 g·mol$^{-1}$ |
| | $\chi_G$ ($^1$H-NMR) | 14 mol% | 18 mol% |
| | $M_n$; $M_w$ (GPC) | 5,200 g·mol$^{-1}$; 6100 g·mol$^{-1}$ | 11,800 g·mol$^{-1}$; 14,700 g·mol$^{-1}$ |
| | PD (= $M_w/M_n$) | 1.17 | 1.25 |
| | $T_g$; $\Delta C_P$ (DSC)$^a$ | -58 °C; 0.11 J·(g·K)$^{-1}$ | +39 °C; 0.56 J·(g·K)$^{-1}$ |
| | $T_m$; $\Delta H_m$ (DSC)$^a$ | +38 °C, +44 °C; 57 J·g$^{-1}$ | n.a. |
| **Network properties** | $Q = 1 + \frac{\rho_2}{\rho_1}\left(\frac{m_s}{m_d} - 1\right)$ (vol%) | 520 ± 10% | 1070 ± 70% |
| | $G = \frac{m_d}{m_{iso}}$ (wt.%) | 99 ± 1% | 97 ± 2% |
| | $T_g$; $\Delta C_P$ (DSC) | -50 °C; 0.43 J·(g·K)$^{-1}$ | +51 °C; 0.47 J·(g·K)$^{-1}$ |
| | $T_m$; $\Delta H_m$ (DSC) | +18 °C; 0.1 J·g$^{-1}$ | n.a. |
| | E (37 °C, aqua) | 3.9 ± 0.0 MPa | 113 ± 13 MPa |
| | $\varepsilon_B$; $\sigma_{max}$ (37 °C, aqua) | 60 ± 20 %; 1.4 ± 0.3 MPa | 355 ± 30%; 11.2 ± 2.4 MPa |

$\chi_G$ = molar fraction of glycolide; $M_n$ = number average molecular weight; $M_w$ = weight average molecular weight; PD = polydispersity; $T_g$ = glass transition temperature; $\Delta C_P$ = changes in heat capacity at $T_g$; $T_m$ = melting temperature; $\Delta H_m$ = melting enthalpy; Q = degree of swelling [mass of swollen ($m_s$) and dried ($m_d$) samples and densities of solvent $\rho_1$ ($\rho_{CHCl3}$ = 1.483 g·cm$^{-3}$) and network $\rho_2$ ($\rho_2$ = 1.215 g·cm$^{-3}$)]; G = gel content [swelling in 100fold excess of CHCl$_3$ for 24 h; $m_{iso}$: mass after synthesis; $m_d$: mass after extraction]; E = elastic modulus; $\varepsilon_B$ = elongation at break; $\sigma_{max}$ = tensile strength; n.a. = not applicable

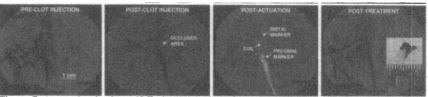

**Figure 7.** As previously shown[26], demonstration of *in vivo* clot extraction using the hybrid SMP-SMA coil device. Rabbit carotid angiograms were acquired pre-clot injection, post-clot injection, during device actuation, and post-treatment. Also pictured is the retrieved clot (scale divisions in mm). Though difficult to visualize here, the entire device was clearly visible on the monitors in the operating room (resolution was lost in transferring the images to video tape

## SMP Stents

We developed two prototypes of the stent in addition to solid wall tubes, which were studied for comparison to the machined stents and for animal implantation studies[21,31,32]. The stent fabrication process included dip coating the thermoset SMPs (both DiAPLEX and LLNL SMPs) and laser cutting desired patterns on a cylindrical mandrel. Our initial stent design was optimized for flexibility (Figure 8). This design also turned out to have hoop stresses comparable to commercial alloy stents of the same diameter. The latter design was laser-deployed in a PDMS model artery with and without flow.

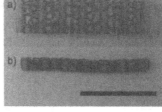

**Figure 8.** As previously shown[32], SMP stent fabricated by laser micromachining. The struts are designed to provide flexibility in both the a) expanded and b) collapsed forms. Scale bar = 1 cm. The outer diameter of the expanded stent is 4.4 mm with a wall thickness of 200 μm.

Compression of the stent pictured in Figure 8 and the solid-walled tubes (not shown) was experimentally studied. The prototype SMP stents used in this study were fabricated from Diaplex MM7520 thermoplastic polyurethane. At 37 °C, both stents exhibited full collapse pressures higher than 4.7 psi, the estimated maximum vasospastic pressure that could collapse an intracranial stent. Full collapse pressure of the laser-machined stent was similar to low collapse pressure stents reported in the literature [33-37]. The stents showed full recovery after crimping, with a radial expansion ratio up to 2.72 and an axial shortening of 1.1 %; higher recovery ratios could be obtained by further dilation.

## SMP Embolic Foams

Recently we have been working on developing SMP embolic foams. The hypothesis is that implanted SMP foams can be used to treat cerbrovascular aneurysms by fully and efficiently filling aneurysms with a clot. The aneurysm is considered non-threatening once the outer clot

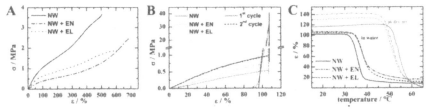

**Fig. 2:** Characterisation of drug-free, EN-, and EL-loaded network (NW). A: Stress-strain curve from tensile test at 70 °C in air. B: Strain-controlled, cyclic thermomechanical experiment in air. C: Shape recovery curves upon heating in a stress-controlled experiment.

deformed shape either under stress- or under strain-control (programming), and finally heating in order to allow the sample to regain its original shape (recovery). For amorphous materials, programming bases on the vitrification of the switching domains in the stretched shape by reducing T below $T_g$. In this context it should be stressed that $T_g$ describes a thermal phase transition determined by DSC, while $T_{sw}$ is related to the macroscopic, thermomechanical properties of a test specimen.

By soaking technique, drug loadings of 2.6 wt.-% (EN) and 1.6 wt.-% (EL) were reached for the N-LG network. Drug crystals were observed by polarized microscopy but, theoretically, drug might also be present in a separate amorphous phase beside the polymer phase or molecularly dispersed in the polymer. In the latter case, plasticizing effects with a reduction in $T_g$ could occur. Since no $T_g$ reduction was found in DSC measurements, the higher $\varepsilon_B$ and lower $\sigma_{max}$ for drug loaded dry samples in tensile tests (Fig. 2A) cannot solely be attributed to plasticisation. Also, a separate amorphous drug phase distributed like islets or in an interpenetrating manner in the amorphous N-LG might be hypothesized to strengthen the material like it is known for nano-composites rather than softening it. Therefore, more studies aiming to reveal drug effects on the thermo-mechanical properties will need to be conducted.

In strain-controlled, cyclic thermomechanical experiments in a dry air atmosphere, congruent curves in ε-σ diagrams were observed for several cycles of programming and recovery for drug-free and drug-loaded samples (Fig. 2B) indicating the preserved integrity of the network architecture during this procedure. $T_{sw}$, being the inflection point of T-ε recovery curves, was reduced due to plasticisation in aqueous environment (Fig. 2C) when changing to aqueous environment. Hydrophobic EN may reduce the initial water uptake and therefore induce a lower shift in $T_{sw}$. Surprisingly, samples loaded with hydrophilic EL showed the same rather than a larger shift of $T_{sw}$ than EN. This indicates that besides water uptake other mechanisms may be involved.

By fitting release profiles by linear regression analysis, rate constants were calculated for different release phases after initial burst of less than 20 wt.-%.

**Table 1:** Kinetic parameter of drug release as obtained by linear regression analysis*.

| Drug | $I_t$ / D | $k_0$ / $10^{-2} \cdot d^{-1}$ | $r^2$ / - |
|------|-----------|-------------------------------|-----------|
|      | 0.1 - 27  | 0.02                          | 0.997     |
| EN   | 27 - 82   | 1.24**                        | 1.000     |
|      | 82 - 310  | 0.07                          | 0.931     |
|      | 0.1 - 26  | 1.23                          | 0.988     |
| EL   | 30 - 45   | 0.43                          | 0.999     |
|      | 52 - 134  | 0.33                          | 0.981     |

$I_t$: studied time interval; $k_0$: release rate constant; $r^2$: square of correlation coefficient; *Release from $10 \cdot 10 \cdot 0.2$ mm$^3$ samples at 37 °C in phosphate buffer; a zero order drug release has been assumed for each of the release phases. ** Due to the small number of data points this value is an approximation only.

Water-soluble EL was released in the first phase at highest rates, while practically no release of hydrophobic EN was observed. Such lag phases are common in release curves of hydrophobic or large molecular weight drugs from bulk-eroding materials, which are caused by limited drug diffusion through the matrix. When the three-dimensional architecture of the polymer network broke down after about four weeks as determined in a loss of gel content (data not shown), EN-release rates increased substantially. After 90 days, more than 85% of both drugs were released. 100% drug release occured after about 130 days for EL and 300 days for hydrophobic EN.

## Conclusion

Two different biodegradable polymer networks from an amorphous respectively a semicrystalline star-shaped telechelic were synthesized in this study. Drug loading and release was achieved for both materials. However, only the crosslinked polyester urethane network from amorphous precursors exhibited the desired shape-memory effect. Overall, this stresses the challenge to avoid reciprocal disturbance of different functionalities in the development of multifunctional materials. Semi-crystalline SMP with star-shaped architecture can potentially be realized by employing precursors with higher $M_n$. Moreover, detailed studies of influences of drug particle incorporation on the thermomechanical properties of SMP are of interest.

## ACKNOWLEDGEMENT

The authors thank for financial support by the BMBF Biofuture grant Nr. 0311867.

## REFERENCES

1. M. Behl, A. Lendlein, Soft Matter 3 (2007), 58-67.
2. K. Gall, C.M. Yakacki, Y. Liu, R. Shandas, N. Willett, K.S. Anseth, J. Biomed. Mater. Res. Part A 73A (2005) 339-348.
3. A. Lendlein, R. Langer, Science 296 (2002) 1673-1676.
4. A.A. Sharp, H.V. Panchawagh, A. Ortega, R. Artale, S. Richardson-Burns, D.S. Finch, K. Gall, R.L. Mahajan, D. Restrepo, J. Neural Eng. 3 (2006) L23-L30.
5. W. Small, T.S. Wilson, P.R. Buckley, W.J. Benett, J.M. Loge, J. Hartman, D.J. Maitland, IEEE Trans Biomed Eng 54 (2007) 1657-1666.
6. Th. Weigel, R Mohr, A Lendlein, Smart Mater. Struct. 18 (2009) 025011 (9pp).
7. R. Mohr, K. Kratz, T. Weigel, M. Lucka-Gabor, M. Moneke, A. Lendlein, Proc. Natl. Acad. Sci. USA 103 (2006) 3540-3545.
8. A. Lendlein, H. Jiang, O. Jünger, R. Langer, Nature 434 (2005) 879-882.
9. A.T. Neffe, D.B. Hanh, S. Steuer, A. Lendlein, Polymer networks combining controlled drug release, biodegradation, and shape-memory capability, Adv. Mat. *in press*, DOI: 10.1002/adma.200802333
10. C. Wischke, A. Neffe, S. Steuer, A. Lendlein. Evaluation of a degradable shape-memory polymer network as matrix for controlled drug release. J. Controlled Release *in press*, doi:10.1016/j.jconrel.2009.05.027.

# Stimuli-sensitive Systems, Actuators, and Sensors

Mater. Res. Soc. Symp. Proc. Vol. 1190 © 2009 Materials Research Society    1190-NN08-01

# Schizophrenic Molecules and Materials With Multiple Personalities - How Materials Science Could Revolutionise How we do Chemical Sensing

*Robert Byrne, Silvia Scaramagnani, Alek Radu, Fernando Benito-Lopez and Dermot Diamond*

CLARITY: Centre for Sensor Web Technologies, National Centre for Sensor Research, School of Chemical Sciences, Dublin City University, Dublin 9, Ireland

## Abstract

Molecular photoswitches like spiropyrans derivatives offer exciting possibilities for the development of analytical platforms incorporating photo-responsive materials for functions such as light-activated guest uptake and release and optical reporting on status (passive form, free active form, guest bound to active form). In particular, these switchable materials hold tremendous promise for microflow-systems, in view of the fact that their behaviour can be controlled and interrogated remotely using light from LEDs, without the need for direct physical contact. We demonstrate the immobilisation of these materials on microbeads which can be incorporated into a microflow system to facilitate photoswitchable guest uptake and release. We also introduce novel hybrid materials based on spiropyrans derivatives grafted onto a polymer backbone which, in the presence of an ionic liquid, produces a gel-like material capable of significant photoactuation behaviour. We demonstrate how this material can be incorporated into microfluidic platforms to produce valve-like structures capable of controlling liquid movement using light.

## Introduction

Chemical sensors are devices that provide information about binding events happening at the interface between a sensitive film/membrane and a sample phase. The function of the sensitive film/membrane is to ensure that the binding at this interface is as selective as possible usually by means of entrapped or covalently bound receptor sites. The binding event is further coupled with a transduction mechanism of some kind; such as a change in the colour or fluorescence of the film or a change in electrochemical potential. Clearly, these materials are 'active' as binding events must occur for them to be of any analytical use. However, it is self-evident that these sensitive interfaces will change over time, for example due non-specific binding and biofouling in real samples that can lead to surface poisoning or occlusion, or leaching of active components into the sample phase. Consequently, the response characteristics of chemical sensors and biosensors will change with time, leading to gradual decrease in sensitivity, loss of selectivity and baseline drift. In practice, these effects are compensated for by regular calibration, until the device deterioration reaches some limiting level. In recent years, physical transducers have been increasingly deployed in sensor networks. However, for equivalent widely distributed chemical sensing to happen, there must be a revolution in the way chemical sensors/biosensors are employed, as conventional calibration is inappropriate for large-scale deployments due to the cost of ownership (particularly maintenance) of these rather complex devices. In this paper, we consider the use of materials that can be switched reversibly between two or more different 'personalities with radically different characteristics. For example, materials that can exist in a passive form (non-binding) until a measurement is required, at which point the material is switched to an 'active' or binding form. Once a measurement is made, the material is switched back to the 'passive' form. This effect may have important potential applications in sensors, purification resins, separation science and drug delivery. We will also suggest how polymer actuators may provide routes to new active components in microfluidic manifolds, such as pumps and valves, that could form the basis of soft-polymer based circulation systems for handling samples, reagents, and standards in futuristic analytical devices that have a distinct biomimetic character.

## Sensornets

'Sensornets' are large-scale distributed sensing networks comprised of many small sensing devices equipped with memory, processors, and short-range wireless communications capabilities.[1] These devices, known as 'Motes' can gather and share sensor data from multiple locations through in-built wireless communications capabilities. The vision of incorporating chemical and biological sensing dimensions into these platforms is very appealing, and the potential applications in areas critical to society are truly revolutionary.[2] For example, the 'environmental nervous system' concept likens the rapid access and response capabilities of widely distributed sensor networks to the human nervous system; i.e. it is able to detect and categorise events as they happen, and organise an appropriate response.[3] Sensors monitoring air and water quality will be able to provide early warning of pollution events arising at industrial plants, landfill sites, reservoirs, and water distribution systems at remote locations.

The crucial missing part in this scenario is the gateway through which these worlds will communicate; how can the digital world sense and respond to changes in the molecular world? Unfortunately, it would appear from the lack of field deployable devices in commercial production that attempts to integrate molecular sensing science into portable devices have failed to bear the fruits promised; this problem is what we call 'the chemo-/ bio-sensing paradox',[4] i.e. *Chemo/bio-sensors **must have an** 'active' or responsive surface incorporating sites that are pre-designed to bind with specific target species in order to generate the chemically or biologically inspired signal; at the same time, these surfaces should be passive, in that they should be resistant to effects that cause signal drift and loss of sensitivity.*

The interactions involved in these binding events can be very subtle, and even slight changes in the surface or bulk characteristics through processes like leaching, fouling, or decomposition, can have a significant effect on the output signal, and the overall performance of the device. This is in contrast to physical transducers, as they can function without having to make direct physical contact with the 'real world'. For example, thermistors are completely enclosed in a tough protective epoxy coating that enables heat to pass through from the real world, and light sensors, which are also completely enclosed, leaving a transparent window through which light can penetrate from the region under observation. When chemo/bio-sensors are exposed to the real world, their sensitive surfaces immediately begin to change, and hence they suffer from baseline drift and variations (usually reduction) in sensitivity, as well as cross-response to interferents that may be present in the sample.

In analytical science, we deal with this issue through regular calibration, meaning that the sensing surface is periodically removed from the sample and exposed to standards, the response characteristics checked, and any baseline drift or change in sensitivity compensated. However, if this type of capability is to be provided to an autonomous chemo/bio-sensing platform, it requires that a liquid flow system is integrated, comprising pumps, valves, and interconnects. This drives up the complexity, price and power requirements of these platforms, and makes the realisation of small, autonomous, reliable, chemical sensing/biosensing devices impractical at present.

Therefore, the traditional vision of the 'chemo/bio-sensor' as a device with an active membrane attached to a pen-like probe is outdated, and needs to be completely rethought. In particular, the issue of how to predict and control surface characteristics at the interface between the device and the real world needs fresh thinking, as this is where the molecular interactions that generate the observed sensor signal happen. The key to progress will require breakthroughs arising from new concepts in fundamental materials science, such as the development of *adaptive* or *stimuli-responsive materials* that have externally or locally controllable characteristics. These materials could be regarded as having capability to switch between several completely different 'personalities' – schizophrenics at the molecular level!

In this paper, we shall show how certain photochromic molecules display intriguing switchable characteristics, and suggest ways in which this can be used to control the function and behaviour of sensing devices and platforms.

# Experimental

## Materials and instruments

(1'-(3-carboxypropyl)-3',3'-dimethyl-6-nitrospiro(2H-1)benzopyran-2, 2'-(2H)-indole) (SPCOOH, Figure 1) was synthesized as reported elsewhere [5]. Polybead carboxylate microbeads 2.035 μm diameter, 2.79% solid contents, were purchased from PolySciences Inc. Plain silica microspheres (5 ± 0.35 μm diameter, 5 % solid contents), N-(3-dimethylaminopropyl)-N'-ethylcarbodiimide hydrochloride (EDC hydrochloride), (3-Aminopropyl)triethoxysilane (APTES), 2-(N-morpholino)ethanesulfonic acid hydrate (MES hydrate), 1,8-diaminooctane, calcium nitrate hydrate, copper(II) nitrate trihydrate and zinc chloride were purchased from Sigma Aldrich (Ireland). Ammonia solution 25% was purchased from Sharlau Chemie (Spain). Homogeneous suspensions of microbeads were generated using a Bransonic Ultrasonic Cleaner 5510 from Branson Ultrasonics Corporation, USA. UV (375 nm), white (430-760 nm), blue (430 nm), green (525 nm), red (630 nm) LEDs were purchased from Roithner Laser Technik, Austria. The UV light source used for the solution studies was a BONDwand UV-365nm obtained from Electrolyte Corporation, USA. Sample spinning was carried out using a ROTOFIX 32 centrifuge (Global Medical Instrumentation, Inc., USA.). Absorbance spectra were recorded using a Well Plate Spectrometer (Medical Supply Co., Ireland). Reflectance spectra were recorded using a miniature diode array spectrophotometer (S2000®) combined with an FCR-7UV200-2 reflection probe (7 X 200 micron cores) and a DH-2000-FSH deuterium halogen light source (215-1700 nm, Ocean Optics Inc., Eerbeek, Netherlands). A white reflectance standard WS-1-SL was used to standardise the measurements at 100% reflectance (Ocean Optics Inc., Eerbeek, Netherlands).

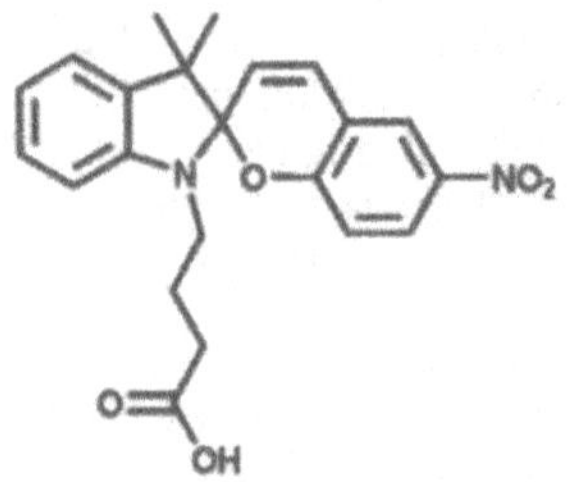

**Figure 1.** 1'-(3-carboxypropyl)-3',3'-dimethyl-6-nitrospiro(2H-1)benzopyran-2, 2'-(2H)-indole (SPCOOH).

## Covalent immobilisation of spiropyran on the surface of silica microspheres

*1) Amino groups coating of the silica microsphere surface*

A sample of a suspension of silica microspheres (0.1 g in 2 ml of water) was diluted with 22.5 ml ethanol. 2.5 ml of ammonia solution 25% and 2 ml of APTES were added and the mixture stirred under reflux for 48 hours. The microspheres suspension was then cooled, separated from the reaction mixture by centrifugation, suspended in 4 ml of ethanol and washed 6 times with fresh ethanol. The washing procedure consists of a four step process:

1. Centrifugation of the suspension for 3 minutes at 4000 rpm.
2. Removal of the supernatant, addition of 4 ml of fresh solvent.
3. Sonication of the suspension for 5 minutes.
4. Subsequent further centrifugation.

*2) Covalent immobilization of Spiropyran on the surface of amino groups functionalised silica microbeads*

3 ml of a 15 mg ml$^{-1}$ solution of SPCOOH in ethanol was added to 2 ml of an 11 mg ml$^{-1}$ solution of EDC in ethanol. The reaction mixture was stirred for half an hour at room temperature in the dark. A 1 ml suspension of 0.1 g of the amino functionalised microspheres in ethanol was added to the spiropyran/EDC solution and the reaction mixture stirred for 72 hours at room temperature in the dark. Finally, the spiropyran functionalised microbeads were profusely washed 10 times with ethanol following the above procedure and stored at 4 °C in the dark.

## Evaluation of metal interactions with SP-coated microspheres

Spiropyran coated silica and polystyrene microspheres suspended in ethanol were exposed to the same concentrations ($10^{-4}$ M) of ethanolic solutions of Ca$^{2+}$ and Cu$^{2+}$ (polystyrene microspheres) and Ca$^{2+}$, Cu$^{2+}$ and Zn$^{2+}$ (silica microspheres) in order to evaluate the complex formation at the bead surface.

Each experiment was carried out using the following procedure:

1. Exposure of the microspheres to a white LED for 1 minute (promotes MC conversion to the SP form)
2. Recording of the spectrum of the colourless SP form.
3. Exposure of the microspheres to a UV LED for 1 minute (promotes SP conversion to the MC form).
4. Recording of the spectrum of the coloured MC form.
5. Addition of $10^{-4}$ M metal solution in ethanol to the microspheres.
6. Recording of the spectrum.

7. Exposure of the microspheres to a white LED for 1 minute (promotes metal expulsion and MC conversion to the SP form
8. Recording of the spectrum of the colourless SP form.

### Preparation of photo-responsive phosphonium based ionogel
The ionogel consists of three monomeric units; N-isopropylacrlamide (NIPAAm), N,N-methylene-bis(acrylamide) (MBAAm) and acrylated benzospiropyran in the ratio 100:5:1, respectively (Scheme 1). The acrylated benzospiropyran is synthesised as described elsewhere [6]. A reaction mixture solution placed in the micro-fluidic reservoir was prepared by dissolving the NIPAAm monomer (452 mg, 4.0 mmol), the MBAAm (10.8 mg, 0.07 mmol), acrylated spirobenzopyran monomer (14.0 mg, 0.04 mmol), and the photo-initiator dimethoxy-phenylacetophenone (DMPA) (10.2 mg, 0.04 mmol) into 1-butanol (1.0 mL). These monomers were photo-polymerised within an ionic liquid matrix. For this example, the ionic liquid matrix used was trihexyltetradecylphosphonium dicyanoamide $[P_{6,6,6,14}][dca]$. UV irradiation source for polymerization (365 nm) was placed 8 cm far from the monomers. When the polymerization was completed, the gels were washed with ethanol and 0.1 mM HCl aqueous solution for 10 min to remove the unpolymerised liquid and the excess of ionic liquid. Finally the ionogels were kept for two hours in 0.1 mM HCl aqueous solution, where the ionogel exhibits a drastic and rapid swelling effect.

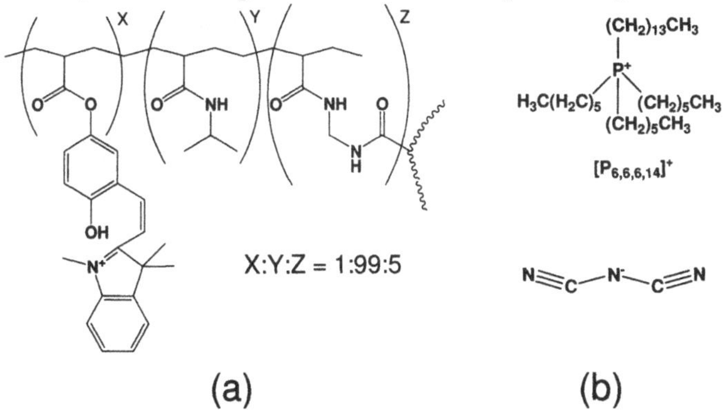

(a)                    (b)

**Scheme 1.** a) Photo-responsive polymer poly(N-isopropylacrylamide), N,N-methylene-bis(acrylamide) and acrylated benzospiropyran in the ratio 100:5:1, respectively. b) trihexyltetradecylphosphonium dicyanoamide $[P_{6,6,6,14}][dca]$ ionic liquid.

### Micro-fluidic device fabrication
The micro-fluidic device shown in Figure 2 (4 x 4 cm) was easily fabricated in poly(methyl methacrylate) (PMMA) and pressure-sensitive adhesive (PSA) in four layers using $CO_2$ ablation laser and it consist of five independent micro-channels. The ionogel valve is placed in a square reservoir (300 × 300 μm), fabricated using the $CO_2$ laser, within the PMMA 125 μm and the PSA 80 μm thickness layers. It is important to mention here that micro-valves can be easily relocated simply by varying the layer layout. A second PSA layer with the channel structures (80 μm deep, 150 μm width, and 20 mm length) was fabricated using the $CO_2$ laser and terminally glued to the

previous PSA layer to generate the micro-fluidic structure. Finally the upper PMMA layer, which contains the inlets and outlets, closes the micro-fluidic structure.

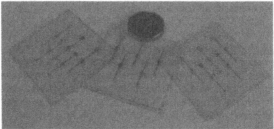

**Figure 2.** Picture of three micro-fluidic devices fabricated in PMMA:PSA polymer by $CO_2$ ablation laser system. Channels were filled with different dyes to improve channel visualization.

## Characterization of the photo-responsive phosphonium based ionogel

Volume phase transition behaviour of the photo-responsive phosphonium based ionogel was investigated using visible light irradiation. The physical shrinking by photo-induced dehydration of the ionogel was measured on-line by contact profilometer experiments when visible light is applied, Figure 3, and by visual observation using a PARISS: "Prism and Reflector Imaging Spectroscopy System" equipped with a CCD camera.

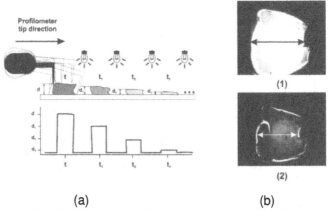

(a)                                         (b)

**Figure 3.** a) Schematic representation of the set-up used for the profilometer measurements. b) Ionogel shrinking process: 1- during white light irradiation, 2- two seconds after white light irradiation, size decrease 29 % in volume.

## Evaluation of micro-valve function

In order to evaluate the function of the ionogel micro-valves fabricated on the micro-fluidic device, a red dye, $B_{12}$ vitamin 1 μM concentration in water was placed into the inlet reservoir. In the outlet, a constant vacuum was applied as the driving force.

144

Visible light irradiation was carried out at 5 mm separation from the valve. When light is applied the polymer decreases in volume and the valve opens, letting the coloured liquid pass through the channel. The liquid movement was observed with a CCD camera.

## Results and Discussion

There is considerable interest in spiropyrans and similar photochromic molecules due to their potential applications in areas such as light-sensitive eyewear,[7] information recording and processing,[8] optical memory,[9] and molecular devices.[10] We are interested in their optical sensing applications [11-14] and in particular their potential for playing multiple roles within a chemical sensor (photoswitch, ligand, transducer). Utilizing photo-switchable molecules such as spiropyrans, which can be switched between active or passive forms, may enable such behaviour to be realised. Upon exposure to UV light, the colourless spiropyran (SP) molecule undergoes a heterocyclic ring cleavage that results in the formation of the merocyanine (MC) form (Scheme 2), which has a deep purple colour as it is a planar, highly conjugated chromophore with a strong absorbance in the visible spectrum. It is well known that, after removal of UV light, the predominant MC form generated will thermally isomerize back to its equilibrium state; and this decay in absorbance typically follows first-order kinetics. [15] Furthermore, the MC isomer has a phenolate anion site to which ions, protons and amino acids can bind, giving rise to new absorption bands in the visible spectrum. [16-18] By shining white light on the colored complex, the guest species is released, and the spiropyran form is regenerated.

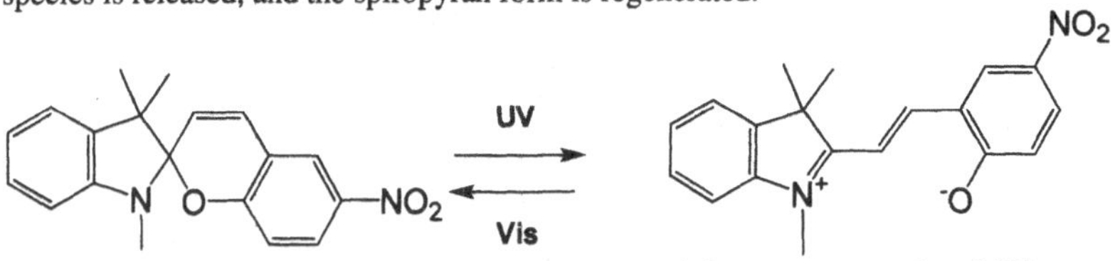

**Scheme 2.** Structures of nitrobenzospiropyran (SP), left, and merocyanine (MC), right.

Therefore these photoswitchable molecules possess some of the characteristics we are interested in employing in next generation analytical platforms. However, it is a significant challenge to manifest these characteristics in a materials format that is compatible with the microfluidic platforms we employ. In order to achieve this, we need to incorporate the switchable moiety within a polymeric matrix through a range of approaches such as physical entrapment within bulk polymers,[19] grafting to monomers,[5] or standard surface immobilisation chemistries like EDC coupling.[20] We are particularly interested in bringing addition innovation to this process, for example, by generating novel hybrid materials such as photoswitchable ionogels (see below).

### Computational analysis
In order to assist with this research, and to better understand the molecular basis for observed characteristics, we have recently employed computational analysis techniques to aid in the design of innovative hybrid materials comprising of spiropyran derivatives and ionic liquids. Standard *ab initio* molecular orbital theory and density functional theory (DFT) calculations were carried out using GAUSSIAN

03 [21], which enabled excitation energies of the MC isomer to be computed at CIS(D)/6-31+G(d,p) level. All molecular orbitals were plotted using the HF/6-31+G(d,p) electron density. The molecular orbitals on the MC isomer are shown in Figure 4. The first excitation arising from the HOMO has an energy of 569 nm, and as this excitation is allowed by symmetry, it is clearly evident in the visible spectrum.

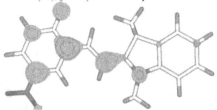

**Figure 4.** HOMO-1 molecular orbital of the MC isomer.

We have also studied the interactions of organic cations and anions with the MC isomer, such as imidazolium [emim]$^+$ and dicyanoamine [dca]$^-$. For [emim]$^+$, we found that the [emim]$^+$ cation appears to interact predominantly via the C2 carbon which resides only 2.92 Å (the C---O distance, Figure 5a) away from the phenolate oxygen. The sum of the van der Waals radii of the carbon and oxygen atoms is 3.22 Å and hence, the C---O separation appears to be considerably shorter than a normal close contact. This C2 hydrogen is in fact known to be quite acidic.[22] The C2 hydrogen atom sits in between the carbon and oxygen atoms (CHO bond angle close to linear) introducing a very short/strong hydrogen bond of just 1.865 Å (H–O distance). As a comparison, a strong hydrogen bond, e.g. O–H--O, is slightly longer in length, around 1.9 Å.[23] Moreover, such a close proximity of the two ions results in a through-space orbital interaction, with molecular orbitals on both oxygen and carbon atoms overlapping, and forming a strong interaction between the two molecules. Due to the through-space interaction, the species formed is likely to be quite stable and will thus hinder conversion of the MC form back to the aplanar SP isomer.[24] The optimised structures of MC with the [dca]$^-$ anion are shown in Figure 5b. This suggests that the two conjugated systems are tilted away from one another, a feature which will also inhibit the thermal relxation of MC back to SP state, as the two conjugated systems have a much lower degree of interaction. These calculations have led us to prepare similar systems and investigate these materials experimentally, as described below.

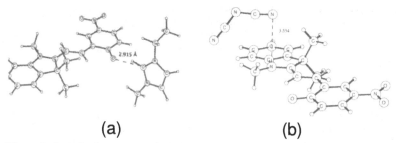

(a)                                    (b)

**Figure 5.** Optimized structures of MC isomer with the (a) [emim]$^+$ and (b) [dca]$^-$ at B3LYP/6-31+G(d).

## Chemical Sensing utilizing bead technology

In the past we have utilized polymeric supports for photo-regenerable chemical sensing applications. [20, 25] Bead based systems are particularly interesting from the chemical sensing perspective as they have higher surface area compared to flat surfaces, which can enhance kinetics compared to conventional surfaces.[26] In addition beads can be moved like a fluid but are easily separated from the liquid phase and hence their incorporation into flow systems for separation science holds many advantages.[27, 28] Recently spiropyran has been covalently immobilised on the surface of silica and polystyrene microspheres and their light-modulated characteristics including ion-binding capability have been investigated.[29] Covalent functionalisation involves the use of a carboxylated spiropyran derivative (Figure 1) which is attached to the bead surface via EDC coupling chemistry.[20] The resulting SP-functionalised microspheres can be switched back and forth between a pink coloured MC form and a colourless SP form using a 375 nm UV-LED (SP⇒MC switching) and a 430-760nm white LED (MC⇒SP switching) (Figure 6).

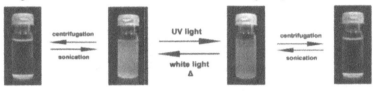

**Figure 6.** Spiropyran switching between the colourless SP and the pink MC form on the surface of the silica functionalised microspheres.

The colour changes have been detected using reflectance spectroscopy (in the case of polystyrene microspheres, due to their high scattering effect) and absorbance spectroscopy (in the case of silica microspheres, due to their higher transparency). Upon irradiation with UV-LED for 1 minute, the pink MC form is generated on the microsphere surface, and the characteristic 560 nm band appears, further exposure of the MC-microspheres to a white LED for 1 minute causes reconversion to the colourless SP form and the disappearance of the characteristic MC band, as seen in Figure 7. When the MC-microspheres are exposed to certain metal ion solutions, they undergo further spectral and visible colour changes, due to the formation of $MC_2\text{-}M^{2+}$ complexes. Subsequently exposure of the beads to illumination with a white LED causes the metal-ion guest to be expelled and the SP form is restored, ready for another ion-binding event. In the case of the silica microspheres the greatest ion binding effects were observed in the presence of $Zn^{2+}$ and $Cu^{2+}$ ions (Figure 7). Under exposure of the MC-microspheres to $Zn^{2+}$, a clear visual colour change from purple (MC) to light pink ($MC_2\text{-}Zn^{2+}$) can be observed. This is caused by a decrease in the MC absorbance band around 560 nm and the emergence of a new absorbance band centred around 525 nm arising from the MC-metal ion complex, consistent with the replacement of free MC by the ion-complex on the microsphere surface.

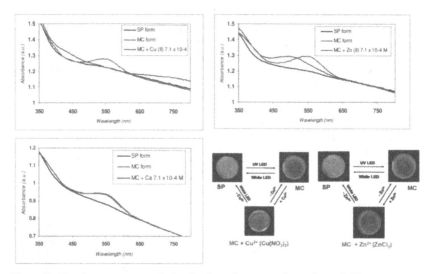

**Figure 7.** Absorbance spectra and visual colour changes on the surface of silica microspheres in the presence of the MC and SP form and after exposure of the MC form to metal ions ($Cu^{2+}$, $Zn^{2+}$ and $Ca^{2+}$).

Similarly, the addition of $Cu^{2+}$ ions to the MC-microsphere sample results in a colour shift from purple to orange (formation of $MC_2$-$Cu^{2+}$ complex), which arises from the disappearance of the 560 nm MC peak and the simultaneous appearance of two new absorbance bands at 440 nm and 750 nm. In both cases, after the formation of the MC-metal ion complex, replacement of the metal ion solution with clean ethanol, followed by irradiation of the microbeads for 4 minutes with a white LED, leads to expulsion of the bound $Cu^{2+}$ ions and complete reformation of the SP form on the beads. Following this, irradiation of the microspheres for 2 minutes with the UV-LED converts the SP back to the MC form, ready for another metal ion uptake and release cycle. This light-modulated ion retention and release behaviour, coupled with visible indication of the bead state, opens the possibility of developing photocontrolled stationary phases that can be activated and deactivated using light. These results show that spiropyran modified beads can be used for the photo-controlled selective accumulation and release of ions. Furthermore, the system is inherently self-indicating, as each form (SP, MC, $MC_2$-$M^{2+}$ complex) has a different colour and UV-vis spectrum. Clearly, this behaviour could have many interesting applications in selective pre-concentration on certain ions, transport of bound ions to remote locations, and controlled release of bound ions, using light as the external controlling stimulus. Figure 8 shows functionalised polystyrene microspheres packed into an optically transparent silica capillary being switched between the two forms (SP and MC). In this form, the beads can clearly be applied to photocontrolled metal-ion uptake and release in a capillary flow system.

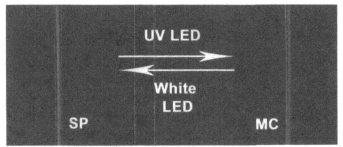

**Figure 8.** Switching of SP-modified polystyrene microspheres between the SP and MC forms in a silica microcapillary using UV and white LEDS.

## Polymeric Actuation using light- Applications in Microvalve Technology.

Even with their impressive physical changes, electro-actuators employing soft polymer materials like conducting polymers have not solved the micro-actuator problem within the field of microfluidics. This is due to the complex fabrication schemes required to incorporate these materials into microfluidic manifolds. Therefore, controlling physical properties with temperature,[6, 30] photon irradiation,[31, 32] or specific chemical (pH, ionic strength)[33, 34] stimulii would be of great benefit for rapid prototyping.

We have developed a photo-responsive microvalve from a hybrid material (ionogel), see Scheme 1. In an acidic aqueous system in the dark, the ionogel (MC-H$^+$) is in a protonated open-ring form. When irradiated by blue light, the MC-H$^+$ isomerizes immediately to a closed-ring form (SP), dissociating protons and losing positive charges. When the light is turned off, the chromophore returns spontaneously to the protonated open-ring form, which is more thermodynamically stable in the dark than the closed-ring form. The photoisomerization affects the hydration of the ionogel significantly. Under dark conditions, the MC-H$^+$ form (hydrophilic) ionogel is yellow. When the ionogel is irradiated with blue light, the ionogel decolorizes, indicating the SP form (hydrophobic) is present. Simultaneously, the gel reduces in size due to loss of charge, and associated induced dehydration of water from the polymer. This photo-induced dehydration results in a physical shrinkage of the ionogel, as seen in the images within Figure 3b. Our results indicate that the photo-induced ionogel shrinkage proceeds through two distinct steps; at first, the isomerization of the MC-H$^+$ to closed-ring SP (uncharged) form take place under white light irradiation for 3 s, calculated experimentally to be 2.5 x10$^{-2}$s$^{-1}$. The resultant hydrophobic isomer induces the dehydration of polymer main chain. After 150 seconds of exposure to visible light, there is a 73 % decrease in the ionogel height, Figure 3a, as monitored using a physical contact profilometer. The rate constant for the slower (shrinkage) stage was experimentally calculated to be 0.457 s$^{-1}$.[35] Optical control of a microvalve structure built into a microfluidic channel was demonstrated as shown in Figure 9. A drop of a solution containing a red dye was placed in the inlet of the channel while at the outlet a vacuum was applied (Figure 9a). The microvalve was irradiated with white light as shown in Figure 9b, and the micro-valve opened after 3 s, allowing liquid to pass trough the channel to the outlet, Figure 9c-e. The light intensity necessary to control the ionogel micro-valve is not particularly intense, for example, a simple white LED ca. 1 mW cm$^{-2}$ [19] can be used to actuate the valve.

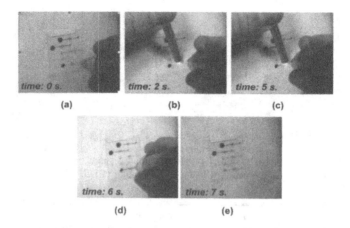

**Figure 9.** Performance of the ionogel micro-fluidic valve. a) Micro-valve is closed; vacuum is unable to draw the red dye through the micro-channel. b) White light is applied opening the micro-valve. c-e) The red dye moves along the micro-channel.

The isomerization of the MC-H$^+$ induced by white light irradiation is a reversible phenomenon. The closed-ring form returns spontaneously to the protonated open-ring form, which induces the swelling of the ionogel. Therefore, it is possible to reuse the micro-valves repeatedly, although the ionogel in its present form requires more than 30 min to swell again and block the whole channel. Nevertheless, a large number of micro-valves can be fabricated simultaneously using the same process described above and therefore, the method is suitable for large-scale integration of micro-valves on microfluidic manifolds. The micro-valve control by light irradiation provides non-contact operation and independent manipulation of multiple fluids on microfluidic devices, as well as parallel control of multiple micro-valves. It is expected that such photoresponsive polymer gel micro-valves will be an advantageous technique for integrated multifunctional micro-fluidic devices.[36]

## Conclusions

The key to many new technologies lies in the development of materials that exhibit stimulus-responsive behaviour. This area has undergone rapid growth in recent years, as the science and technology of molecular and nano-scale control and characterization of materials continues to develop. The range of materials that can be switched between dramatically different modes of behaviour is expanding rapidly, and in this paper, we have only been able to provide an introduction into some of the exciting possibilities that can arise from materials containing spiropyran derivatives. Even within the limits of this particular paper, it is clear that these materials could revolutionise the capabilities of analytical platforms, facilitating light modulated uptake and release of molecular guests on channel surfaces and beads, transport of bound species to other locations on beads, photo-switchable separation of sample components, light-actuated valves and pumps to control liquid flow, which, if integrated with simple optical detection, could provide a route to low-cost analytical platforms with whose characteristics are completely controlled using light. And beyond the analytical world, these materials have the potential to be incorporated into a wide range of specialist and consumer products within the next 5 years that could dramatically impact on society. Furnishings and clothes that change colour, textiles that can sense and communicate, chemical sensors whose surface binding activity can be turned on/off, and materials with light switchable hydrophobicity/permeability. All in all, it seems clear that there are exciting times ahead in sensor science aligned with adaptive or stimuli-responsive materials!

## Acknowledgements

Authors would like to thank Science Foundation Ireland for continued support, under grants 07/CE/L1147 (CLARITY) and 07/RFP/MASF812. FBL would like to thank the Irish Research Council for Science, Engineering and Technology (IRCSET) fellowship number 2089. RB would like to thank Dr. Ekaterina Izgorodina for her helpful discussions on theoretical calculations.

## References

1.  Shenker S, et al., ACM SIGCOMM Computer Communication Review, 2003. **33**(1): p. 137-142.
2.  Diamond, D., *Internet-scale sensing*. Analytical Chemistry, 2004. **76**(15): p. 278A-286A.
3.  Diamond, D., *Internet-scale chemical sensing: is it more than a vision?* NATO Security through Science, Series A: Chemistry and Biology, 2006. **2**(Advances in Sensing with Security Applications): p. 121-146.
4.  Byrne, R. and D. Diamond, *Chemo/bio-sensor networks*. Nature Materials, 2006. **5**(6): p. 421-424.
5.  Rosario, R., et al., *Photon-Modulated Wettability Changes on Spiropyran-Coated Surfaces*. Langmuir, 2002. **18**(21): p. 8062-8069.
6.  Szilagyi, A., et al., *Rewritable Microrelief Formation on Photoresponsive Hydrogel Layers*. Chemistry of Materials, 2007. **19**(11): p. 2730-2732.
7.  Crano, J.C., et al., *Photochromic compounds: chemistry and application in ophthalmic lenses*. Pure and Applied Chemistry, 1996. **68**(7): p. 1395-1398.
8.  Guglielmetti, R., *Spiropyrans and related compounds [applications]*. Studies in Organic Chemistry (Amsterdam), 1990. **40**(Photochromism: Mol. Syst.): p. 855-78.
9.  Dvornikov, A.S. and P.M. Rentzepis, *Accessing 3D memory information by means of nonlinear absorption*. Optics Communications, 1995. **119**(3,4): p. 341-6.
10. Willner, I., et al., *Reversible light-stimulated activation and deactivation of a-chymotrypsin by its immobilization in photoisomerizable copolymers*. Journal of the American Chemical Society, 1993. **115**(19): p. 8690-4.
11. Collins, G.E., et al., *Photoinduced switching of metal complexation by quinolinospiropyranindolines in polar solvents*. Chemical Communications (Cambridge), 1999(4): p. 321-322.
12. Collins, G.E., et al., *Spectrophotometric Detection of Trace Copper Levels in Jet Fuel*. Energy & Fuels, 2002. **16**(5): p. 1054-1058.

13.  Winkler, J.D., C.M. Bowen, and V. Michelet, *Photodynamic Fluorescent Metal Ion Sensors with Parts per Billion Sensitivity.* Journal of the American Chemical Society, 1998. **120**(13): p. 3237-3242.

14.  Suzuki, T., et al., *Photo-reversible Pb2+-complexation of insoluble poly(spiropyran methacrylate-co-perfluorohydroxy methacrylate) in polar solvents.* Chemical Communications (Cambridge, United Kingdom), 2003(16): p. 2004-2005.

15.  Gorner, H., *Photochromism of nitrospiropyrans: effects of structure, solvent and temperature.* Physical Chemistry Chemical Physics, 2001. **3**(3): p. 416-423.

16.  Gorner, H. and A.K. Chibisov, *Complexes of spiropyran-derived merocyanines with metal ions - Thermally activated and light-induced processes.* Journal of the Chemical Society-Faraday Transactions, 1998. **94**(17): p. 2557-2564.

17.  Evans, L., III, et al., *Selective Metals Determination with a Photoreversible Spirobenzopyran.* Analytical Chemistry, 1999. **71**(23): p. 5322-5327.

18.  Ipe, B.I., S. Mahima, and K.G. Thomas, *Light-Induced Modulation of Self-Assembly on Spiropyran-Capped Gold Nanoparticles: A Potential System for the Controlled Release of Amino Acid Derivatives.* Journal of the American Chemical Society, 2003. **125**(24): p. 7174-7175.

19.  Stitzel, S., R. Byrne, and D. Diamond, *LED switching of spiropyran-doped polymer films.* Journal of Materials Science, 2006. **41**(18): p. 5841-5844.

20.  Byrne, R.J., S.E. Stitzel, and D. Diamond, *Photoregenerable surface with potential for optical sensing.* Journal of Materials Chemistry, 2006. **16**(14): p. 1332-1337.

21.  Frisch, M.J.T., G. W.; Schlegel, H. B.; Scuseria, G. E.; Robb, M. A.; Cheeseman, J. R.; Montgomery, Jr., J. A.; Vreven, T.; Kudin, K. N.; Burant, J. C.; Millam, J. M.; Iyengar, S. S.; Tomasi, J.; Barone, V.; Mennucci, B.; Cossi, M.; Scalmani, G.; Rega, N.; Petersson, G. A.; Nakatsuji, H.; Hada, M.; Ehara, M.; Toyota, K.; Fukuda, R.; Hasegawa, J.; Ishida, M.; Nakajima, T.; Honda, Y.; Kitao, O.; Nakai, H.; Klene, M.; Li, X.; Knox, J. E.; Hratchian, H. P.; Cross, J. B.; Bakken, V.; Adamo, C.; Jaramillo, J.; Gomperts, R.; Stratmann, R. E.; Yazyev, O.; Austin, A. J.; Cammi, R.; Pomelli, C.; Ochterski, J. W.; Ayala, P. Y.; Morokuma, K.; Voth, G. A.; Salvador, P.; Dannenberg, J. J.; Zakrzewski, V. G.; Dapprich, S.; Daniels, A. D.; Strain, M. C.; Farkas, O.; Malick, D. K.; Rabuck, A. D.; Raghavachari, K.; Foresman, J. B.; Ortiz, J. V.; Cui, Q.; Baboul, A. G.; Clifford, S.; Cioslowski, J.; Stefanov, B. B.; Liu, G.; Liashenko, A.; Piskorz, P.; Komaromi, I.; Martin, R. L.; Fox, D. J.; Keith, T.; Al-Laham, M. A.; Peng, C. Y.; Nanayakkara, A.; Challacombe, M.; Gill, P. M. W.; Johnson, B.; Chen, W.; Wong, M. W.; Gonzalez, C.; and Pople, J. A., *GAUSSIAN 03.* 2004, Gaussian Inc: Wallingford CT.

22.  MacFarlane, D.R., et al., *Lewis base ionic liquids.* Chemical Communications (Cambridge, United Kingdom), 2006(18): p. 1905-1917.

23.  Modig, K., B.G. Pfrommer, and B. Halle, *Temperature-Dependent Hydrogen-Bond Geometry in Liquid Water.* Physical Review Letters, 2003. **90**(7): p. 075502.

24.  Byrne, R., et al., *Photo- and solvatochromic properties of nitrobenzospiropyran in ionic liquids containing the [NTf2]- anion.* Physical Chemistry Chemical Physics, 2008. **10**(38): p. 5919-5924.

25. Radu, A., et al., *Photonic modulation of surface properties: a novel concept in chemical sensing.* Journal of Physics D: Applied Physics, 2007. **40**(23): p. 7238-7244.

26. M.-S. Kim, et al., Transducers, 2003.

27. Adam, T., Lüdtke S., and K. K. Unger, *Application of 0.5-μm porous silanized silica beads in electrochromatography* Journal of Chromatography A, 1997. **786**(2): p. 229-235.

28. Adam, T., Lüdtke S., and K. K. Unger, *Packing and stationary phase design for capillary electroendosmotic chromatography (CEC).* Chromatographia, 1999. **49**: p. S49-S55.

29. Silvia Scarmagnani, et al., *Polystyrene bead-based system for optical sensing using spiropyran photoswitches.* Journal of Materials Chemistry, 2008. **18**: p. 5063-5071.

30. Reber, N., et al., *Transport properties of thermo-responsive ion track membranes.* Journal of Membrane Science, 2001. **193**(1): p. 49-58.

31. Sugiura, S., et al., *Photoresponsive polymer gel microvalves controlled by local light irradiation.* Sensors and Actuators, A: Physical, 2007. **A140**(2): p. 176-184.

32. Kameda, M., et al., *Photoresponse gas permeability of azobenzene-functionalized glassy polymer films.* Journal of Applied Polymer Science, 2003. **88**(8): p. 2068-2072.

33. Kim, S.J., et al., *Surprising shrinkage of expanding gels under an external load.* Nature Materials, 2006. **5**(1): p. 48-51.

34. Eddington, D.T., et al., *An organic self-regulating microfluidic system.* Lab on a Chip, 2001. **1**(2): p. 96-99.

35. Sugiura, S., et al., *Photoresponsive polymer gel microvalves controlled by local light irradiation.* Sens. Actuators, A FIELD Full Journal Title:Sensors and Actuators, A: Physical, 2007. **A140**(2): p. 176-184.

36. Sugiura, S., et al., *On-demand microfluidic control by micropatterned light irradiation of a photoresponsive hydrogel sheet.* Lab Chip, 2009. **9**: p. 196-198.

Mater. Res. Soc. Symp. Proc. Vol. 1190 © 2009 Materials Research Society          1190-NN08-02

## Color Switchable Goggle Lens Based on Electrochromic Polymer Devices

Chao Ma and Chunye Xu[1]
[1]Center of Intelligent Materials and Systems, University of Washington, Box 352600, Seattle, WA 98115, E-Mail:chunye@u.washington.edu

## ABSTRACT

We have developed a set of new electrochromic devices (ECDs) for special application goggles, whose color can be switched between transparent and a specific color mode, i.e. blue (B). This paper will discuss the design, film deposition, device assembly and characterizations of the color switchable lens. The ECD is composed of a layer of thin film conducting polymer poly (3,4-(2,2-dimethylpropylenedioxy)thiophene) (PProDOT-Me$_2$), a layer of thin film inorganic oxide V$_2$O$_5$-TiO$_2$, and a layer of ionic conductive electrolyte. The thin films are electrochemically deposited on ITO coated flexible plastic substrate. The whole device is packaged with an UV cured flexible film sealant. The goggle lens exhibit tuneable shade in visible light wave length (380-800nm), with a maximum contrast ratio at 580nm. Meanwhile, other unique properties include fast switching speed, low driving voltage, memory function (no power needed after switching, bistable), great durability, high flexibility, light weight, and inexpensiveness.

## INTRODUCTION

Today people have known more and more about the sun's harmful effects on the eyes. Long exposure to sun light may cause diseases such as cataracts, macular degeneration, and photokeratitis [1, 2]. Mean while unexpected glare, flash and varied light conditions are always safety issues faced by people activating outdoors, including motorcycle riders, truck drivers, and even soldiers. A smart, color switchable eyewear is highly demanded here, which can exhibit different light transmittance instantly upon user's command or changing environment.

The requirement has been beyond the ability of traditional materials for sunglasses or goggles. They have only one fixed color state and the transparency is uncontrollable. Existing color switchable eyewear products on the market, manufactured by Uvex Inc. or Corning Inc., are based on liquid crystal devices (LCDs), or photochromic devices [3-6]. LCDs are capable of switching quickly and perform reasonable high light transmission range, but their disadvantages are also obvious: high operation voltage, complex manufacturing, lack of memory function, and limited color availability. Photochromic devices rely on specific chemical reaction and are activated by UV light. However, photochromic materials based eyewear only reacts to UV irradiation not to visible light. One awkward example is when you sit in a car. Because the wind shield blocks out most of the UV light, photochromic devices will not work inside the car. In addition, the color switching speed is too slow to satisfy fast pace activities.

Electrochromic (EC) materials can change their optical properties reversibly for an applied potential due to electrochemical oxidation and reduction. Electrochromic devices (ECDs) which are able to actively control light transmittance/absorbance could be used in this field to overcome the issues. In our lab, conjugated, conductive polymers have been developed as the powerful EC materials for various applications in terms of their flexibility, light weight, inexpensiveness, and easiness of scalability [7, 8]. This advanced technology can be characterized as quick response, adjustable transmittance, rich color, low driving voltage, memory function (no power needed after switching, bistable), and great durability. This electrochromic device technology can be tailored for color switchable goggles where high contrast ratio and fast switching speed are required. In the presented work, a cathodic blue color EC polymer material, poly (3,4-(2,2-dimethylpropylenedioxy)thiophene) (PProDOT-Me$_2$), structure shown in Figure 1, is utilized to fabricated ECDs.

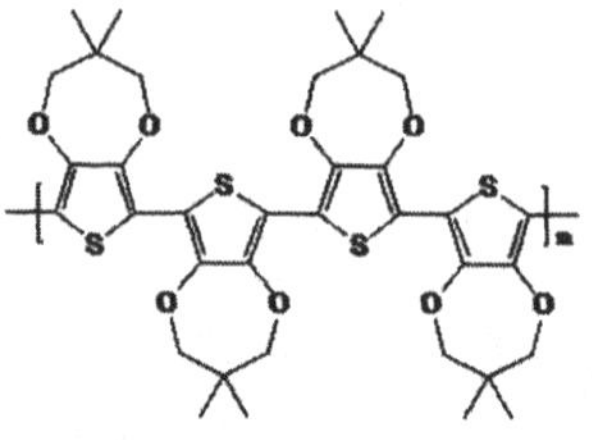

**Figure 1.** Structure of PProDOT-Me$_2$

## EXPERIMENT
### Materials and reagents

ProDOT-Me$_2$ monomer was synthesized in our lab via an improved route. The method was reported by Xu et al. [9, 10]. All starting materials were purchased from Aldrich, except lithium perchlorate (LiClO$_4$, 99% anhydrous, packed under argon), which was purchased from Alfa Aesar. Because the electrochromic polymer film is sensitive to moisture and oxygen, which could affect the performance of EC devices, all the materials were dried before use and stored in glove box filled with argon.

A solution of electrolyte was prepared by dissolving 0.1M of LiClO$_4$ in propylene carbonate (PC). This solution electrolyte was injected into the electrochromic device later using vacuum filling. Special attention must be paid to avoid any contaminations from moisture and oxygen. The solution electrolyte needs to be bubbled by argon gas before use, and all the containers should be dried in oven.

### Working and counter electrode

The PProDOT-Me$_2$ polymer film was deposited from 0.01M monomer in a 0.1M LiClO$_4$/Acetonitrile (ACN) solution. Electrochemical deposition method was carried out on an electrochemical analyzer (CHI 604C, CH Instruments), utilizing the chronoamperometry method. A three electrode cell with Ag wire as a reference, ITO/PET (purchased from CPFilm, 50Ω/ε surface resistance) as a working electrode and a stainless steel plate as a counter electrode was used for electrochemical polymerizing the polymer film.

Vanadium pentoxide and titanium oxide (V$_2$O$_5$/TiO$_2$) composite film was adopted in this study. As the counter film in EC device, the composite film exhibits improved performance over

$V_2O_5$ film used previously. The $V_2O_5/TiO_2$ composite film was deposited on ITO/PET substrate by the chronoamperometry method in $V_2O_5 \cdot TiO_2 \cdot nH_2O$ sol-gel solution. The sols of $V_2O_5 \cdot TiO_2 \cdot nH_2O$ were synthesized using a method reported in previous literature [11].

Both of the EC film and $V_2O_5/TiO_2$ composite film need to be placed into a 0.1M $LiClO_4/PC$ electrolyte solution after polymerization. The films were electrochemically conditioned by the chronocoulometry method in order to change the inorganic ions in the films and have them to be familiar with a $LiClO_4/PC$ environment.

## EC device assembly

A $V_2O_5/TiO_2$ composite film coated ITO/PET was placed on the top of an EC film coated ITO/PET, and the two were clamped together. Solution type electrolyte was injected into the EC device using a vacuum filling system. An UV cured film sealant (IS 90453, obtained from Adhesives Research, Inc.) was used as a hermetic barrier to seal the device.

## Characterization methods

The electrochromic devices were switched using a potentiostat CHI 604C, CH Instruments. Optical characterization of the devices was carried out using an UV-Visible-IR spectrophotometer V-570, JASCO.

## DISCUSSION
### Device design and assembly

EC device based on flexible ITO/PET substrate includes plastic substrate, flexible film sealant and solution type electrolyte, as shown in Figure 2. They exhibit improved flexibility and high electrochromic performance. In this design, the electrochromic working layer, PProDOT-$Me_2$ film, is deposited on indium tin oxide (ITO) coated PET plastic. Because electrochemical polymerization method is adopted here, the deposition area could be patterned by using a mask. The counter layer of the device is vanadium oxide titanium oxide ($V_2O_5/TiO_2$) composite film, also deposited on ITO/PET. The $V_2O_5/TiO_2$ composite film serves as an ion storage layer and works with the PProDOT-$Me_2$ film as a pair. When the EC film is oxidized and doped by $ClO_4^-$ with an applied potential and changes to transparent state, the $V_2O_5/TiO_2$ composite film will absorb $Li^+$ simultaneously. When the EC film is reduced back with an opposite potential and changes to blue color, the $V_2O_5/TiO_2$ composite film will releases $Li^+$ to maintain the charge balance. During switching, the $V_2O_5/TiO_2$ composite film maintains a pale yellow/green color. A transparent solution type electrolyte is sandwiched between the working and counter layers. It is a good conductor for small ions such as $ClO_4^-$ and $Li^+$ and an insulator for electrons. It serves as an ion transport layer and ions move quickly inside during switching.

A pressure adhesive UV curable film sealant is adopted in the flexible EC device assembly. It is sandwiched between working and counter electrodes, and maintains a 30 μm gap between the electrodes; meanwhile block all the moisture and oxygen. Namely, the film sealant is able to play both roles of spacer and barrier. After laminating working and counter electrodes together, the solution type electrolyte is filled into the gap between two electrodes through the inlet port in a vacuum filling system [12], which is specially designed and developed in our lab. The inlet port is sealed by UV cured glue in the final step.

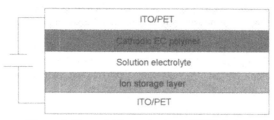

**Figure 2.** The multi layer structure of ECD design

### EC film and counter film deposition

In current states, there is no maturely developed all-plastic electrochromic device based on EC polymers, due to some main technical challenges. One of the unsolved issues is related to the low quality of transparent conductor layer (typically ITO) on plastic substrate, compared with glass substrate. The ITO coating is required to be highly uniform, highly conductive, and highly transparent. Because electrochemical deposition method is used in this study to prepare EC polymer and $V_2O_5/TiO_2$ films, the quality of ITO layer will affect the uniformity of both films and performance of devices. A high quality ITO coated PET plastic was adopted in this study as the substrate. The surface resistance is $50\Omega/\varepsilon$, and light transmittance is 85% at 580nm wavelength.

PProDOT-Me$_2$ polymer film is fragile and sensitive to experimental conditions. Therefore parameters of electrochemical deposition are carefully controlled. Uniform EC film deposition was achieved on the ITO/PET substrate. Since the surface resistance of the substrate is higher than typical glass substrate, potential drop across the substrate surface will be more obvious. Considering this issue, the deposition potential was increased during the study, from 1.5V to 2.5V. When the deposition potential was 2.5V, as shown in Figure 3, high quality, uniform blue EC film coating was achieved. Figure 3 shows the deposited EC film in color and transparent state, respectively. The film was defect free with vivid blue color. The film deposition duration time here is 20 seconds. A copper tape was applied to minimize the potential drop through the substrate surface.

Similarly, the $V_2O_5/TiO_2$ composite film is deposited onto the ITO coated substrate via chronoamperometry method. After wet chemical processing, excess liquid needs to be removed from the deposited film: a baking at over 100°C is required here. Since PET plastic can experience some deformation during heating, it is necessary to control the baking time and temperature. After coating, the plastic substrate is put on a flat substrate and heated at 104°C for 4h. A layer of uniformly deposited counter film on the ITO/PET substrate is shown in Figure 4. The deposition potential is controlled at 3V for 50 seconds.

According to the function of $V_2O_5/TiO_2$ counter film, there are two requirements for it: high uniformity and enough charge capacity. To achieve the charge balance and redox of EC film during color switching, the charge capacity of counter film needs to be equal to or larger than that of the EC film. Figure 5 is the charge plots of counter film and EC film with 1.2V reduction/oxidation potential for 8 seconds. It is clear that the charge plots of EC film are lower than that of the counter film.

(a)                              (b)

**Figure 3**. Uniform blue EC film deposited on the CPFilm substrate in (a) color and (b) transparent state

**Figure 4**. Uniform counter film deposited on the CPFilm substrate

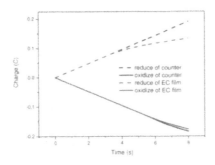

**Figure 5**. Electrical charge plots of counter and EC film during reduction and oxidation

## Light transmittance and color switching speed

After assembly, the EC device was attached to a typical goggle lens, because of its good flexibility. The goggle lens was color switched by a DC potential, and exhibited blue color and transparent state, as shown in Figure 6. The applied potentials were ±1.5V, and charging time was 4s. PProDOT-Me$_2$ polymer film is a cathodic EC material. Therefore, when a -1.5V potential is applied to the working electrode of the lens, the device change its color to dark blue. The color change is due to the reduction of PProDOT-Me$_2$ polymer film. Once the applied potential is put back to +1.5V, the lens will be oxidized and change to transparent state.

Complete color switching of the lens could be achieved by the low driving potentials. The light transmittance of the device was measured and shown in Figure 7. The shown data was picked up at 580nm wavelength, because the lens exhibits maximum contrast ratio ΔT% here. The EC device light transmittance contrast ratio (ΔT%) is defined in equation (1),

159

$$\Delta T\% = T^t(\lambda) - T^c(\lambda) \tag{1}$$

where $T^t(\lambda)$ and $T^c(\lambda)$ is the light transmittance of certain wave length $\lambda$ on transparent state and colored state. It is worth to note that human eyes are especially sensitive to light in 550-600nm region, making the color switching extremely sensitive and effective visibly. The goggle lens exhibited satisfactory $\Delta T\%$. The maximum contrast ratio obtained reaches 48% (18% in color state and 66% in transparent state).

(a)           (b)

**Figure 6**. Lens in (a) colored state and (b) transparent state respectively

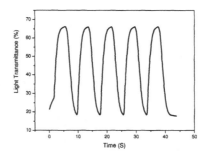

**Figure 7**. Light transmittance of the goggle lens measured at 580nm wavelength

The device was subjected to a cyclic repeated ±1.5V potential, and the color switching was continuous. Charging time was set to 4s for every period. Because during outdoor activities the goggle needs to have fast response to various lights conditions and performs rapid color switching. Thus the color switching speed of the lens is important. The color switching time $T_{90\%}$ is defined as the time at which the optical transmittance reaches above 90% of its full state, timing from either fully colored or transparent state at 580nm wavelength. As observed in Figure 7, $T_{90\%}$=3s for both bleaching and coloring the goggle lens. This color switching speed is fast for an all-plastic EC device of that size, and is quite enough for hiking, skiing or motorcycling and daily driving. The lens also performed good repeatability and stability. During continues switching between colored and transparent state, the optical spectrum curves of different cycles are stable and almost identical to each other. There is no obvious decay or aging as time passing.

## CONCLUSIONS

In this study, a set of new electrochromic devices (ECDs) for special application goggles was designed and fabricated. It was based on ITO coated PET flexible substrate. The EC working part was PProDOT-Me$_2$ polymer film, while the counter part was V$_2$O$_5$/TiO$_2$ composite film. Solution type electrolyte was used as an ion transportation layer. The goggle lens was sustained and sealed by a transparent UV cured film sealant. The developed plastic lens exhibited adjustable transmittance of light (18%-66%), fast response time (about 3s), low driving potential (1.5V) and good repeatability. Meanwhile it had its own advantages in more flexibility, less weight and volume, and easier patterning. This flexible plastic based ECDs could be widely used in manufacturing color switchable, smart goggles.

## ACKNOWLEDGMENTS

Financial support from Revision Eyewear Ltd. is gratefully acknowledged.

## REFERENCES

1. S.J. Dain, *Clin Exp Optom.* **86**(2), 77 (2003).
2. P. Kullavanijaya and H.W. Lim, *J Am Acad Dermatol.* **52**(6), 937 (2005).
3. C. Xu, L. Liu, S.E. Legenski, D. Ning, and M. Taya, *J. Mater. Res.* **19**(7), 2072 (2004).
4. T.A. Skotheim, R.L. Elsenbaumer, and J.R. Reynolds, *Handbook of conducting polymers*, 2$^{nd}$ Ed., (Marcel Dekker Inc., New York, US, 1998).
5. D.J. Kerko, D.W. Morgan, and D.L. Morse, U.S. Patent, No 4 608 349 (1986).
6. T.G. Havens, D.J. Kerko, U.S. Patent, No 4 979 976, (1990).
7. C. Ma, M. Taya, and C. Xu, *Electrochimica Acta*, **54**, 598-605 (2008)
8. C. Ma, M. Taya, and C. Xu, *Polymer Engineering and Science*, **48** (11), 2224-2228 (2008).
9. C. Xu, H. Tamagawa, M. Uchida, and M. Taya, *Proceeding of SPIE.* **4695**, 442 (2002).
10. D.M. Welsh, A. Kumar, E.W. Meijer, and J.R. Reynolds, *Advanced Materials.* **11**(16), 1379 (1999).
11. S. Kim, M. Taya, and C. Xu, *Materials Research Society Symposium Proceedings*, **928**, 160-167 (2006).
12. C. Xu, C. Ma, X. Kong, and M. Taya, *Polymers for Advanced Technology*, **20**, 178–182 (2009)

Mater. Res. Soc. Symp. Proc. Vol. 1190 © 2009 Materials Research Society                     1190-NN08-03

Michael Wegener[1], Robert Schwerdtner[2], Martin Schueller[3], and Andreas Morschhauser[2]

[1] Functional Materials and Devices, Fraunhofer Institute for Applied Polymer Research (IAP), Geiselbergstraße 69, 14476 Potsdam, Germany
[2] ZfM – Center for Microtechnologies, Chemnitz University of Technology, Reichenhainer Straße 70, 09126 Chemnitz, Germany
[3] Multi Device Integration, Fraunhofer Research Institution for Electronic Nano Systems (ENAS), Reichenhainer Straße 88, 09126 Chemnitz, Germany

## ABSTRACT

In this work, dome structures in PVDF films were prepared as ultrasonic transducer. The domes are realized by a deep drawing process. The dome-forming process leads to a phase transformation from the non-polar α into the polar β phase within dome wall and roof of the PVDF films. A ferroelectric polarization is obtained in these dome areas after suitable electrical poling which yields a piezoelectric activity. Because of the piezoelectric activity within the film plane, a dome-roof up-and-down actuation is observed with resonance frequencies in the range between 65 and 93 kHz.

## INTRODUCTION

Piezoelectric materials transform mechanical stresses into electrical signals (direct effect, sensor function) or an electrical stimulus into a form change of the original geometry (inverse effect, actuator function). Piezoelectric sensors are used in various applications for measuring direct or air-transferred impacts from low up to ultrasonic frequencies. As actuators, piezoelectric materials deliver motions also at high frequencies and are often embedded as active materials in ultrasonic systems. The actuation amplitude of piezoelectric polymers (as well as of ceramics) are relatively small in comparison e.g. to that of other electro-active polymers (EAP) such as dielectric elastomer actuators (DEA). However, piezoelectric polymers have also advantages in comparison to other EAP's, because they don't need high operating voltages, special environments and frame constructions. Frequently studied piezoelectric polymers are ferroelectric systems such as polyvinylidene fluoride (PVDF) and its copolymers [1, 2] as well as ferroelectrets such as voided space-charge systems made of polyester and polyolefines [3-5].

Ferroelectric PVDF and P(VDF-TrFE) polymers exhibit piezoelectric 33- and 31-coefficients of about ten to some ten pC/N or pm/V. This piezoelectric activity leads to thickness changes of some nm if thin film transducers were operated in the piezoelectric thickness-extension actuator mode. In order to overcome the relatively low actuation in the thickness-extension mode other transducer geometries e.g. using the piezoelectric 31-effect in dome-like or rainbow structures are necessary [6, 7]. The usage of the piezoelectric 31 effect as actuator mode in such polymer domes has two advantages: (i) the expected larger elongation as well as (ii) the adjustment of piezoelectric resonances in the range of 20 to 100 kHz due to the much lower frequency of the length- in comparison to the thickness-extension resonance [8].

Here, we discuss the preparation of dome-like structures in PVDF films by a developed deep drawing process. In detail, we describe the phase transformation in the dome walls and roofs, necessary procedures for the metallization of inner and outer dome surfaces as well as the electrical poling process. Finally, the transformation of piezoelectric actuation from the length extension mode in the dome walls into a change of the dome heights is determined via an interferometer.

**EXPERIMENT**

Deep drawing of dome structures is performed by applying a low pressure of 10 psi ("negative pressure") to the PVDF film while heating up the film with an infrared lamp to about 120°C. Therefore a sample holder containing different amounts of holes with different diameters (1, 2 and 3 mm) connected to a pressure regulator was constructed in order to realize separate samples with either 1, 4 or 9 domes at a single film area. A photograph of the sample holder is shown in Figure 2.

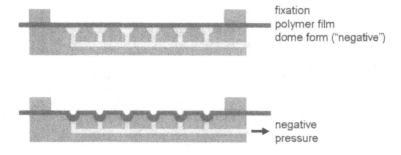

**Figure 1:** Schematic sketch of dome-forming procedures.

**Figure 2:** Sample holder for the attachment of PVDF films during the deep-drawing process.

Figure 3 shows 9-dome structures, a schematic sketch (pictures (a) and (b)) and an experimentally realized system (c). The developed preparation process leads to dome structures with some millimeter diameter of the dome base area and dome heights between 1 and 3 mm without destruction (such as holes and rips) of the PVDF films. Beside the application development of nine-dome structures as ultrasonic transducers, all structural investigation as well as the characterization of the electrical charging behavior and the actuator properties are performed on single domes (as shown in Figure 3(d)) and discussed in the next paragraph.

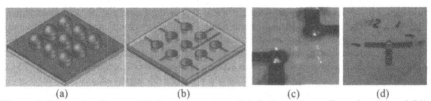

(a)          (b)          (c)          (d)

**Figure 3:** Schematic pictures of 9-dome structures and their electrode configuration ((a) and (b)). 9-dome (c) and 1-dome structure (d) prepared in PVDF films by deep drawing and metallized with copper (2 domes in picture (c)) and aluminum electrodes (d) on both surfaces.

The chain conformations in the PVDF domes were characterized with an FT-IR spectrophotometer (Thermo Nicolet Nexus 470). The spectra were recorded in transmission mode at different positions of the dome. One measurement is performed through the top (roof) of the dome, another through the dome wall. In this case, the infrared beam transmits the sample twice; the IR beam enters the inner dome part through one dome wall and penetrates once again through the dome wall at the opposite side of the dome. In addition, a reference spectrum is recorded at the remaining PVDF film well outside the area of the formed dome.

For electrical poling the dome inner and outer surfaces were metallized with electrodes with a thickness of about 60 nm by thermal evaporation either copper or aluminum. Due to the cylindrical shape of the dome wall, a special sample holder was constructed allowing the shift of the axis of the dome-wall cylinder which is otherwise vertical aligned in comparison to the metallization sources. Thus, a complete metallization of the inner and outer dome wall is possible and demonstrated exemplarily on two domes in Figure 3(c) as well as on the single dome shown in Figure 3(d).

In order to polarize the PVDF domes, combinations of bipolar and uni-polar high-voltage cycles were applied to the samples. The HV device (TREK 610C) is thereby controlled via a function generator (Agilent 33250a). The poling current through each sample was measured in situ by means of an electrometer (Agilent 3458a) operated in the current mode. Current contributions from the charging of the sample capacitance and the conductivity were subtracted from the measured poling currents according to the procedure described in references [9] and [10]. The polarization is obtained by time-integration of the separated polarization current. The advantage of this procedure is the possibility to determine also small polarization phenomena. Those are usually not recognized if the poling current (or charge) is measured during the application of bipolar (e.g. sinusoidal) electric-field – time dependencies. Finally, the piezoelectric actuator properties were characterized by measuring deflection of electrically stimulated samples using laser-doppler-interferometry. The ultrasonic resonance frequencies and the amplitudes of the up-and-down movement of the dome-like structures were determined taking into account the transference function of the laser-doppler-interferometry at the resonance frequency.

## DISCUSSION

As well known from the basic research performed on PVDF, a phase transformation into the polar β phase can be achieved by different procedures such as drawing, high-pressure treatment and/or application of high polarization fields. The phase transformation during these processes will be significantly enhanced if these processing steps will be performed at high temperatures. The finally received β phase exhibit an all-trans chain conformation, a parallel alignment of neighboring chains and thus electrically addressable dipoles with a dipole moment of 2.1 D. The phases are usually characterized by infrared spectroscopic experiments. Different vibrations can be characterized in order to evaluate the amount of α und β phases in the PVDF polymer films [11]. Very common is the characterization by means of a so-called $D_{530}/D_{510}$ coefficient which is calculated from the absorptions at wave numbers of about 530 cm$^{-1}$ (α phase, vibration at 532 cm$^{-1}$) and 510 cm$^{-1}$ (β phase). This coefficient is typically between 0.01 and 12 corresponding to films mostly in the β or in the α phase, respectively.

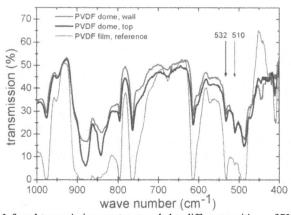

**Figure 4:** Infrared-transmission spectra recorded at different positions of PVDF domes.

The infrared spectra recorded at different positions of a PVDF dome are plotted in Figure 4. The spectrum measured at a reference position in the film plane outside the dome area corresponds to that of non-polar PVDF which consists mostly in its α phase. The dome forming process leads to a phase transformation, as e. g. demonstrated with the increase of the absorptions at 510 cm$^{-1}$. From the calculated $D_{530}/D_{510}$ coefficients (dome wall: 1.1, dome roof 1.3) a slightly higher transformation into the polar β phase is concluded at the dome wall in comparison to the dome roof. That is expected due to the slightly stronger stretching in the wall areas during the deep drawing process. However, in total only a slight phase transformation is obtained with the currently used dome-forming parameters. This follows from a comparison of the here determined $D_{530}/D_{510}$ coefficients with that of a highly polar β-PVDF sample which is typically 0.7 after hot drawing with a stretch ratio of 1:4.

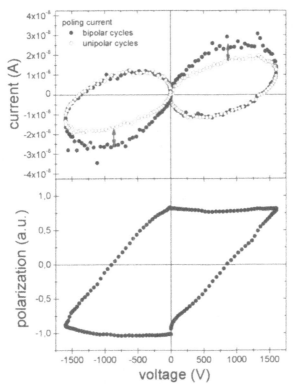

**Figure 5:** Electrical poling of a metallized PVDF dome: Poling current (top) and polarization (bottom) as function of the applied voltage.

The measured poling current during bipolar and uni-polar cycles of the electric field shows a significant difference as indicated in Figure 5 with the arrows at positive and negative poling fields. The values in Figure 5 are given as current and voltage instead of current density and electric field, respectively, because of the not exactly determinable film thickness in the dome walls and the not exactly known metallized dome area. Based on the relatively high frequency of the applied poling field (10 mHz) a significant space-charge effect is excluded. Thus the difference between the uni-polar and bipolar cycles is attributed to a ferroelectric polarization. The minimum remanent polarization is approximately 5 mC/m$^2$ as calculated from an assumed complete metallized inner and outer dome area. This corresponds to the lowest estimated polarization, the value will be enhanced if the dome metallization is not complete which is difficult to analyze. Therefore, the dome shows at least about 15% of the polarization of a well polarized pure β PVDF film. This is an understandable and expected result if the above discussed only slight transformation into the polar β phase is taken into account.

The up to know characterized three dome-like PVDF actuators exhibit resonance frequencies of 65, 73 and 93 kHz for the up-and-down movement of the dome roof. A maximum actuation of 1.5 nm/V is determined for the dome-roof movement at the resonance frequency. A detailed study concerning the dependence of the resonance frequency and maximum actuation on the dome-forming parameters and thus on the thickness and phase transformation as well as on the polarization build up during electrical poling is in progress.

## CONCLUSIONS

In summary, it is demonstrated that dome-like structures are achieved in PVDF polymer films by a developed deep-drawing technique. A destruction of the polymer film could be prevented. The dome forming leads to a transformation from the non polar $\alpha$ into the polar $\beta$ phase in the roof and walls of the PVDF dome. Therefore, a ferroelectric polarization is achieved in the dome area by electrical poling. The designed actuator geometry leads to a successful transformation from a longitudinal in-plane into a thickness movement (of the dome roof) at relatively low resonance frequencies.

## REFERENCES

[1]  T. Furukawa, Structure and properties of ferroelectric polymers, Key Engineering Materials 92-93, 15-30 (1994).
[2]  H.S. Nalwa (Ed.), Ferroelectric polymers, Marcel Dekker Inc., New York (1995).
[3]  M. Wegener and S. Bauer, Microstorms in cellular polymers: A route to soft piezoelectric transducer materials with engineered macroscopic-dipoles, ChemPhysChem 6, 1014-1025 (2005).
[4]  S. Bauer, Piezo-, pyro- and ferroelectrets: Soft transducer materials for electromechanical energy conversion, IEEE Trans. Dielectr. Electr. Insul. 13 (5), 953-962 (2006).
[5]  M. Wegener, Ferroelectrets – Cellular piezoelectric polymers, in: New Materials for Microscale Sensors and Actuators – An Engineering Review, Edited by S. Wilson and C. Bowen, Materials Science and Engineering: R: Reports 56, 78-83 (2007).
[6]  H. Daßler, Mikromechanische realisierte PVDF-Ultraschallwandler-Arrays für Anwendungen in Flüssigkeiten, PhD-thesis (in German), Chemnitz University of Technology (2002).
[7]  H.-J. Kim, H. Lee, and B. Ziaie, A wideband PVDF-on-silicon ultrasonic transducer array with microspheres embedded low melting temperature alloy backing, Biomed Microdevices 9, 83-90 (2007).
[8]  H. Ohigashi, Electromechanical properties of polarized polyvinylidene fluoride films as studied by the piezoelectric resonance method, J. Appl. Phys. 47 (3), 949-955 (1976).
[9]  B. Dickens, E. Balizer, A.S. DeReggi, and S.C. Roth, Hysteresis measurements of remanent polarization and coercive field in polymers, J. Appl. Phys. 72 (9), 4258-4264 (1992).
[10] M. Wegener, Polarization – electric field hysteresis of ferroelectric PVDF films: Comparison of different measurement regimes, Rev. Sci. Instr. 79, 106103 (2008).
[11] M. Kobayashi, K. Tashiro, and H. Tadokoro, Molecular vibrations of three crystal forms of poly(vinylidene fluoride), Macromolecules 8, 158-170 (1975).

Mater. Res. Soc. Symp. Proc. Vol. 1190 © 2009 Materials Research Society 1190-NN11-01

## Experimental and Theoretical Investigation of Photosensitive

## ITO/PEDOT:PSS/MEH-PPV/Al Detector

Leon R. Pinto[1], Jovana Petrovic[2], Petar Matavulj[2], David K. Chambers[1], Difei Qi[1], and Sandra Zivanovic Selmic[1]

[1]Institute for Micromanufacturing and Electrical Engineering Program, Louisiana Tech University, Ruston, LA 71272, USA
[2]Faculty of Electrical Engineering, University of Belgrade, Belgrade 11000, Republic of Serbia

## ABSTRACT

One of widely investigated materials for photodiode, light-emitting device, and solar cell applications is a soluble conjugated polymer poly(2-methoxy-5- (2,9-ethyl-hexyloxy)-1,4-phenylene vinylene) or MEH-PPV. In this paper we present experimental results on MEH-PPV polymer and ITO/PEDOT:PSS/MEH-PPV/Al photodetector, where ITO and PEDOT:PSS stand for indium tin oxide and poly(3,4-ethylenedioxythiophene):poly(styrenesulfonate), respectively. Thin polymer films were fabricated by spin-coating technique. The characterization of the material and devices are done in air at room temperature. The experimental results include optical absorption of MEH-PPV and determination of the optical absorption coefficient, photocurrent dependence on optical power, light wavelength, bias voltage, and polymer thin film thickness. Theoretical modeling is based on drift-diffusion and continuity equations for hole polarons, as well as assumption that the charge carrier recombination process is bimolecular. The bimolecular recombination mechanism implies that the photocurrent depends on the square root of the optical power, which is confirmed with our experimental results.

## INTRODUCTION

Applications of conjugated polymers in electronic devices such as photodetectors [1], solar cells [2], field-effect transistors [3], and light emitting devices [3] are already on their way to be commercialized. This opens up a new era of cheap fabrication and low capital investment in manufacturing of electronic devices. A large number of semiconductive polymers are optically active in the ultraviolet and visible light wavelength range. They are inexpensive and simple to fabricate. However, there are drawbacks in device life time and poor efficiency when compared to device made of inorganic materials such as silicon and gallium. Various studies by researchers [4-7] around the world have shown that there is still more to understand the working mechanism of these devices in areas of carrier generation, transport, and recombination processes.

MEH-PPV poly [2-methoxy-5-(2-ethylhexyloxy)-p phenylvinylene] is a derivative of PPV, which is completely soluble and provides good interest in studying the electron-hole recombination process. In this paper, we present experimental results on MEH-PPV and ITO/PEDOT:PSS/MEH-PPV/Al photodetectors with 6 different MEH-PPV thin film thicknesses, where ITO and PEDOT:PSS stand for indium tin oxide and poly(3,4-ethylenedioxythiophene): poly(styrenesulfonate), respectively. Devices were tested with the monochromator light at room temperature with optical power within the range 1-100 $\mu$W.

Our previous theoretical work includes a model for the ITO/PEDOT:PSS/MEHPPV/Al photodetector based on the solution of the continuity and drift-diffusion equation for hole polarons [7]. We assumed the monomolecular hole polaron recombination mechanism [7]. Our new experimental results presented here indicate that the hole polaron recombination mechanism is bimolecular. Thus, we improved our previous theoretical model by assuming that the recombination for hole polarons is bimolecular in nature and here we present new numerical results for 42nm MEH-PPV thin film thickness.

## EXPERIMENTAL DETAILS

The MEH-PPV (molecular wt. = 260,000 g/mol) was purchased from American Dye Source® and dissolved in chlorobenzene $C_6H_5Cl$ at a concentration of 10 mg/ml in a vial stirred on a hot plate overnight at 90°C. Glass substrates coated with 150 nm ITO with a sheet resistance of 5-15 $\Omega/\square$ from Delta Technologies® were used as the anodic contact for the photodetectors. A water soluble hole transporting polymer poly(3,4-ethylenedioxythiophene): poly(styrenesulfonate) (PEDOT:PSS) was used as a hole transporting material between ITO and MEH-PPV polymer layer. The MEH-PPV solution was filtered using 0.5 µm filters and spincoated over PEDOT-PPS for specified thicknesses. The substrates were later dried in a vacuum chamber overnight and aluminum was thermally deposited for the cathode contacts defining 6 devices. Al wires were attached to each using silver epoxy from Epotek® with a volume resistivity of 0.01 $\Omega$-cm at 23°C with 24 h curing time.

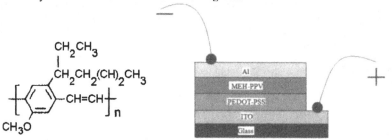

**Figure 1.** Chemical structure of MEH-PPV and layout of the photodetector on ITO coated glass substrates.

Optical power was recorded for the Oriel® 100 W quartz tungsten halogen lamp using a ThermoOriel® silicon diode and Oriel® 70310 optical meter. One device was tested at a time with changing the bias voltage from +1 V to -8 V with 100 ms time for each 0.1 V step. The light wavelength sweep was from 800 nm to 370 nm with10 nm step. Filters with light transmittance of 50 %, 20 %, 10 % and 1 % were used and the photocurrent was recorded for each of them using the Keithly® 6487 voltage source/ picoammeter.

## THEORY

Theoretical modeling is based on drift-diffusion and continuity equations for hole polarons, as well as the assumption that the charge carrier recombination process is bimolecular. The bimolecular recombination mechanism implies that the photocurrent depends on the square root of the optical power:

$$\frac{dp}{dt} = T\theta\alpha I_0 \exp(-\alpha x) - \gamma p^2 - \frac{1}{e}\frac{dJ_p}{dx} \qquad (1)$$

where $p$ is the hole polaron concentration, $T$ is the average transmittance of the ITO/PEDOT:PSS electrode on the glass substrate over the visible wavelength range, $I_0$ is the incident photon flux density, $\theta$ is the internal quantum efficiency of charge generation by light, $\alpha$ is absorption coefficient of MEH-PPV, and $p^2\gamma$ is a hole polaron bimolecular recombination rate. The last term is a gradient of a hole polaron current density $J_p$ given by the drift-diffusion equation:

$$J_p = e\mu_p pE - eD_p\frac{dp}{dx} \qquad (2)$$

where $e$ is the elementary charge, $E$ is the total electric field in the device, $\mu_p$ and $D_p$ the hole polaron mobility and the diffusion coefficient, respectively. We used Poole-Frenkel expression for hole polaron mobility dependence on $E$:

$$\mu_p = \mu_0 \exp\left(\sqrt{E/E_0}\right) \qquad (3)$$

where $\mu_0$ is zero-field mobility and $E_0$ is a characteristic field.

## RESULTS AND DISCUSSION

In pristine polymer photodetectors, the exciton (bound electron-hole pair) dissociation happens in the vicinity of the electrodes or other dissociation centers like defects. The exciton diffusion length for PPV-based polymers is relatively short ~10 nm [4]. Thus, the active region for photogeneration of excitons that can contribute to the current generation is only 10 nm from the polymer/electrode or polymer/polymer interface [5] and thin polymer films are preferred. On the other hand, charge carrier generation from excitons can be attained not only on interfaces and dissociation centers if high electric field (~$10^6$ V/cm) is applied to polymer film and if the film is photoexcited with high incident photon flux (~$10^{15}$ 1/cm$^2$s) [8-9]. In this case strong exciton annihilation processes occur in which charge carriers, namely hole and electron polarons, are created [8]. The quantum efficiency of charge carrier photogeneration in polymer film depends on incident photon flux and electric field intensity [7]. Thirdly, thicker MEH-PPV films absorb more photons. Altogether, the wavelength of the photocurrent peak can be expected to shift with polymer film thickness changes because of the trade-off regarding the film thickness (see Figure 2). The red shift of the photocurrent spectral maximum can be explained in the light of the device antibatic response. For thicker films antibatic feature is more pronounced than for thinner films [10].

The photocurrent density dependence on optical power for our devices is shown in Figure 3. The shape of the curve is square root which implies that the charge carrier recombination mechanism is bimolecular. When we include this assumption, our model gives much better agreement between the measured and simulated photocurrent density spectra (Fig. 4) than our previous model that assumed monomolecular hole polaron recombination [7].

171

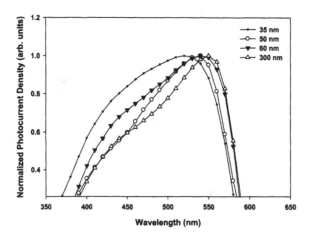

**Figure 2.** The 21 nm red shift of the photocurrent density peak wavelength when active MEH-PPV film thickness changes from 35 nm to 300 nm. The reverse bias voltage for this experiment was -6 V.

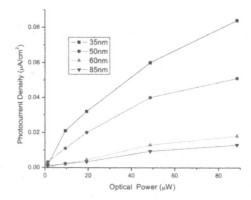

**Figure 3.** Photocurrent density versus optical power for 35nm, 50nm, 60nm, and 85nm thick MEH-PPV thin film based photodetectors for -4 V reverse bias, and illumination with monochromatic 550 nm wavelength light.

172

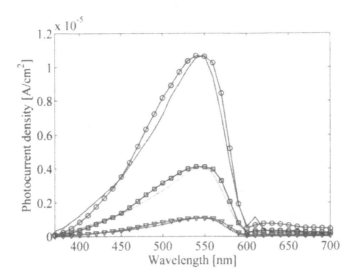

**Figure 4.** Spectral photocurrent density of ITO/PEDOT:PSS/MEH-PPV photodetector for 42nm MEH-PPV thickness. Dashed, dotted, and dash-dotted curves represent the experimental curves for bias voltages of -7V,-5V and -3V, respectively. Curves with circles, squares, and triangles represent the simulated theoretical curves for the case of bimolecular recombination for bias voltages of -7V,-5V and -3V, respectively.

## CONCLUSION

The photocurrent density versus optical power characteristic confirms that the recombination is a bimolecular process. The influence of thickness of active polymer MEH-PPV layer on photocurrent density spectrum is studied. Optoelectronic ITO/PEDOT:PSS/MEH-PPV/Al device characteristics as well as optical characteristics of MEH-PPV film may be useful for better design of photovoltaic devices in the future.

## ACKNOWLEDGMENTS

This material is based upon work in part supported by the Louisiana Experimental Program to Stimulate Competitive Research (EPSCoR), funded by the National Science Foundation and the Board of regents Support Fund contract No. NSF(2008)-PFUND-111, and the Serbian Ministry of Science and Technological Development, contract No. 16001A.

## REFERENCES

1. M.G. Harrison, and J. Gruner, "Analysis of the photocurrent spectra of MEH-PPV polymer photodiodes", *Physical Review B*, v.55, n.12, pp.7831-7849,1997.

2. B.A.Gregg, and M.C Hanna, "Comparing organic to inorganic photovoltaic cells: Theory, experiment, and simulation", *Journal of Applied Physics,* vol. 93, no.6, pp. 3605-3614,2003.

3. M.S Lee, H.S Kang, H.S Kang, J Joo, A.J Epstein, and J.Y Lee ,"Flexible all-polymer field effect transistors with optical transparency using electrically conducting polymers", *Thin Solid Films*, vol.477, no.1-2, pp. 169-173, 2005.

4. J. J. M. Halls and R. H. Friend, The photovoltaic effect in a PPV/perylene heterojunction, *Synthetic Metals*, vol.85, no.1-3, pp.1307-1308, 1996.

5. E.Moons, "Conjugated polymer blends: linking film morphology to performance of light emitting diodes and photodiodes", *Journal of Physics: Condensed Matte,* vo.14, pp.12235-12260, 2002.

6. A.J. Lewis, A. Ruseckas, O.P.M. Gaudin, G.R. Webster, P.L Burn, I.D.W.Samuel, "Singlet exciton diffusion in MEH:PPV films studied by exciton-exciton annihilation," *Organic Electronics*, vol.7, pp.452-456, 2006.

7. J. Petrović, P. Matavulj, D. Qi, D. K. Chambers, and S. Šelmić, "A Model for the Current-Voltage Characteristics of ITO/PEDOT:PSS/MEHPPV/AL Photodetectors," *IEEE Photonics Technology Letters*, vol.20, no.5, pp.348-350, March 2008.

8. B. Kraabel, V.I. Klimov, R. Kohlman, S. Xu, H-L. Wang, D.W. McBranch, ''Unified picture of the photoexcitations in phenylene-based conjugated polymers: Universal spectral and dynamical features in subpicosecond transient absorption," *Phys. Rev. B*, vol. 61, pp. 8501-8515, 2000.

9. V. I. Arkhipov, H. Bassler, M. Deussen, E.O. Gobel, ''Field-induced exciton breaking in conjugated polymers," *Phys. Rev. B,* vol. 52, pp. 4932-4940, 1995.

10. H. B. DeVore, "Spectral Distribution of photoconductivity," *Physica Review*, vol. 102, no. 1, pp. 86-91, April 1956.

Mater. Res. Soc. Symp. Proc. Vol. 1190 © 2009 Materials Research Society          1190-NN11-04

Studies of the blue to red phase transition in polydiacetylene nanocomposites and blends

Anitha Patlolla[1], Qi Wang[2], Anatoly Frenkel[2], James L. Zunino III[3], Donald R. Skelton[3] and

Zafar Iqbal[1*]

[1]Department of Chemistry and Environmental Science, New Jersey Institute of Technology, Newark, NJ 07102
[2]Department of Physics, Yeshiva University, New York, NY 10033 and Brookhaven National Laboratory, Upton, NY 11973
[3]U.Ş. Army ARDEC, Picatinny Arsenal, NJ 07806

## ABSTRACT

The conjugated polymeric backbone of polydiacetylenes (PDAs), comprising of alternating ene-yne groups, undergo intriguing stress-, chemical- or temperature-induced chromatic phase transitions associated with the disruption of the backbone structure and shortening of the conjugation length. PDAs, such as polymerized 10, 12 pentacosadiynoic acids (PCDA), when incorporated with inorganic oxides form nanocomposites and uniform blends with polymers. Blends of poly-PCDA with polymers, such as polyvinyl alcohol, polyvinylidene fluoride and cellulose increase the blue to red transition temperature without affecting the irreversibility of the red phase. However, the addition of zinc oxide to pure poly-PCDA makes the red phase highly reversible and substantially increases the blue to red transition temperature. The addition of $TiO_2$ to poly-PCDA on the other hand does not affect the irreversibility of the red phase and the chromatic transition temperature. In order to understand the atomic scale interactions associated with these changes in the chromatic transitions, we have investigated both the nanocomposites and polymer blends using Raman and Fourier-transform infrared spectroscopy, and extended X-ray absorption fine structure (EXAFS) measurements.

## Introduction

Polydiacetylenes (PDAs) are a unique class of conjugated polymers which undergo dramatic thermochromic and chemically-induced chromatic transitions [1] due to shortening of the conjugation length of the polymeric backbone [2]. Modification of the polydiacetylenes as nanocomposites integrated into inorganic host matrices can have significant effects on their chromatic transitions [3]. It was also reported that the thermochromic transition temperatures are a function of the nature of the side groups on the polymer backbone and the structure of the substituents [4]. A number of the polydiacetylenes in the form of nanocomposites have been synthesized and their chromatic responses to various stimuli have been demonstrated.

Previous studies of the mechanism of the solid state phase transitions in the polydiacetylenes [5] have demonstrated that structural changes occur independently in the polydiacetylene backbone and hydrocarbon side chains of PDAs to give rise to the chromatic transitions. It was also reported that changes do not occur directly in the distribution of electron density along the backbone when chromatic effects are induced [6], but side chain interactions lead to strain on the diacetylene backbone [7] to give rise to the chromatic transformations. Many of the thermochromic transitions are irreversible with decrease of temperature, but chemically-induced chromatic transitions which enhance multiple hydrogen bonding, strong aromatic or ionic interactions between head groups or some level of covalent bonding, can be reversed by removing the adsorbed molecules to recover the original length of the conjugated $\pi$ electron backbone [8].

In this work, near-resonance Raman and Fourier transform infrared (FTIR) spectroscopy together with EXAFS (Extended X-ray Absorption Fine Structure) measurements have been used to study PDA-polymer and PDA-metal oxide interactions, and the possible nature of the chromatic phase transitions in these materials. The PDA used in the present study is poly (10, 12-pentacosadynoic acid).

## EXPERIMENTAL DETAILS

Zinc oxide, titanium dioxide, cellulose, and polyvinyl alcohol (PVA) were purchased from Sigma-Aldrich, PCDA (10, 12-pentacosadiynoic acid) was purchased from GFS Chemicals, and polyvinyldene fluoride (PVDF) was purchased from Polysciences Inc. The average molecular weights of PVA are 30,000-70,000 from low angle light scattering and of PVDF is 80,000. The polydispersity values for PVDF and PVA are: ~2.5-3.0 and ~1.9, respectively. These parameters are not available for cellulose.

1:1 ratio by weight poly-PCDA - metal oxide composites investigated in this study were prepared by mixing solutions and suspensions in chloroform of PCDA and the oxides of titanium and zinc. The solids obtained were UV irradiated at 254 nm for 5 minutes to form the polymeric poly-PCDA blue phase and heated above the transition temperature to form the red phase. 1:1 ratio by weight polymer blends of poly-PCDA with PVA, PVDF and cellulose were similarly prepared and converted to the blue and red phases. Thin films from the solutions or suspensions were prepared by dip-coating on silicon wafers at room temperature. The chromatic transition temperatures were measured on thin films prepared by spin coating using an aliquot of 2 mM of PCDA in chloroform on microscope glass slides at the rate of 2000 rpm. Spin coating films of different metal oxide suspensions containing 1:1 weight ratio of PCDA and metal oxide and polymer suspensions, was performed at the same rate on glass slides.

Powders of poly-PCDA-metal oxide composites in the blue and red phases were used for the EXAFS measurements. Raman measurements were performed both on thin films and powders of the nanocomposites and polymer blends. Samples for FTIR spectroscopy were prepared by pressing KBr pellets of the nanocomposites and polymer blends in a die.

FTIR spectra were obtained using a Perkin Elmer FTIR spectrometer. The Raman spectra were obtained using a Mesophotonics Raman spectrometer with 785 nm laser excitation. EXAFS measurements were carried out using the Synchrotron Light Source at the Brookhaven National Laboratory. SEM and optical images were obtained with a LEO scanning electron microscope and a Ken-A-Vision optical imaging microscope.

## RESULTS AND DISCUSSION

Scanning electron microscope (SEM) images of zinc oxide and titanium dioxide powders are shown in Fig. 1 (a) and Fig. 1 (b), respectively. A relatively wide distribution of sizes above and below 100 nm is seen for zinc oxide, whereas titanium dioxide shows a more uniform distribution of particle sizes around 100 nm. An optical image of a poly-PCDA: PVDF blend shown in Fig. 1(c) indicates a fairly uniform mixing of the two polymers at a spatial resolution of around 2 μm.

The chromatic blue to red phase transitions are of three types: reversible, partially reversible, and irreversible, which have been explained in terms of hydrogen bonding interactions.

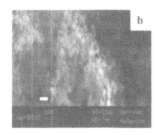

**FIGURE 1.** (a) and (b) Scanning electron microscope (SEM) images with scale bar shown below the images of pure zinc oxide powder and pure titanium dioxide powder, respectively; and (c) Optical microscope image of 1:1 weight ratio of polydiacetylene (poly-PCDA): PVDF taken with a 40x objective at a spatial resolution of approximately 2 μm.

Complete thermochromic reversibility from the red to the blue phase takes place in PDAs where sufficiently strong hydrogen bonding interactions remain unchanged throughout the thermal cycle [9]. Interactions of this type involving double hydrogen bonding among the head groups can also be induced by the addition of specific organic molecules. It was therefore surprising that the addition of the inorganic compound ZnO to the PDAs induced chromatic reversibility and a large upshift of the chromatic transition temperature. By contrast, the addition of $TiO_2$ and $ZrO_2$ did not affect the chromatic transition parameters.

The FTIR spectrum for the poly-PCDA-ZnO nanocomposite in Fig. 2c shows a large increase in intensity of the line at 1540 cm$^{-1}$. This line is relatively weak in the FTIR spectrum of pure poly-PCDA (Fig. 2a) and is not observed in the spectrum of the nanocomposite of poly-PCDA with $TiO_2$ (Fig. 2b). The 1540 cm$^{-1}$ line is also observed with relatively high intensity in the FTIR spectrum of the red phase of the poly-PCDA-ZnO nanocomposite (not shown). However, by contrast to the FTIR spectrum in Fig. 2c for the blue phase of the poly-PCDA-ZnO nanocomposite, there is a substantial decrease in intensity of the C=O stretching line at 1688 cm$^{-1}$ in the red phase. This suggests an interaction between ZnO and the -COOH head group of the PDA side chain, resulting in infrared activation of the line at 1540 cm$^{-1}$, which is close to the Raman frequency at 1515 cm$^{-1}$ of the C=C stretching mode in the red phase (Fig. 4b). The line at 1540 cm$^{-1}$ can therefore be assigned to an infrared-activated C=C stretching mode in the poly-PCDA-ZnO nanocomposite. The corresponding C=C stretching resonance-enhanced Raman line in the blue phase is at 1451 cm$^{-1}$ (Fig.3) due to increased conjugation. The infrared activated C=C mode frequency is the same in both the blue and red phases because, unlike the resonance Raman spectra, the conjugation length of the polymer backbone does not influence the infrared spectra. A new Raman line in the C≡C region at 2258 cm$^{-1}$ appears in the red phase but not in blue phase (Fig. 4). Further studies are needed to understand the activation mechanism for this line, which probably also results from changes in the PDA backbone due to the interaction of ZnO with the -COOH group on the side-chain. Other lines appearing in the Raman spectra of the red phase of the poly-PCDA-ZnO nanocomposite in Fig. 4b are due to the emergence of the blue phase in the presence of ZnO. The thermochromic transition of the blue to the red phase in the poly-PCDA-ZnO nanocomposite results in an upshift of the resonantly enhanced C=C stretching Raman frequency from 1449 cm⁻¹ to 1515 cm⁻¹ and of the C≡C stretching Raman frequency from 2089 cm⁻¹ to 2122 cm⁻¹ (Fig. 4) due to decrease in conjugation length of the PDA backbone.

The FTIR spectra of poly-PCDA show small differences relative to the pure diacetylene on addition of PVA, PVDF and cellulose, indicating weak interactions between the polymers in these blends. This is consistent with up shifts observed in the transition temperatures of the blends relative to that of the pure polydiacetylene. Preliminary EXAFS studies on poly-PDCA metal oxide nanocomposites show changes in the mean square disorder in Zn-Zn distances in the ZnO-poly-PCDA complex relative to pure ZnO (Fig. 5a), consistent with the bonding interactions between ZnO and poly-PCDA indicated by the FTIR spectra. Similar differences in the EXAFS data are not observed in the poly-PCDA-TiO₂ nanocomposite relative to the data for pure TiO₂ (Fig. 5b).

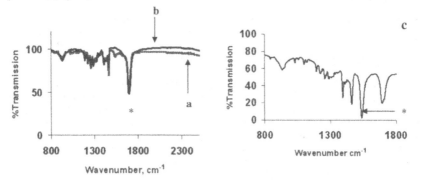

**FIGURE 2.** FTIR spectra of poly-PCDA in the blue phase mixed as composites with metal oxides: Spectrum a) Pure poly-PCDA, b) Poly-PCDA: titanium dixode (1:1 weight %) nanocomposite, and c) Poly-PCDA: zinc oxide (1:1 weight %) nanocomposite. The line at 1540 cm⁻¹ discussed in the text is indicated by an asterisk.

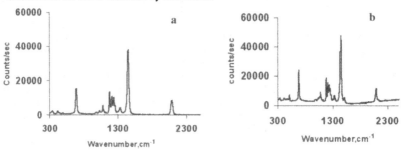

**FIGURE 3.** Near-resonance Raman spectra in the 300 to 2300 cm$^{-1}$ region of: a) Pure poly-PCDA in the blue phase, and b) 1:1 weight % nanocomposite of poly-PCDA with zinc oxide in the blue phase. The spectra were taken with the powder spread uniformly on a silicon wafer.

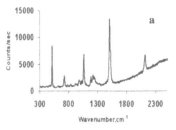
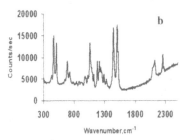

**FIGURE 4:** Near-resonance Raman spectra in the 300 to 2300 cm$^{-1}$ region of: (a) Poly-PCDA in the red phase, and (b) 1:1 weight % nanocomposite of poly-PCDA nanocomposite with zinc oxide in the red phase partially converted to the blue phase.

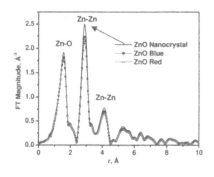
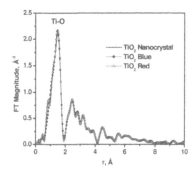

**(a)**                                                                 **(b)**

**FIGURE 5.** EXAFS spectra shown as plots of the Fourier transform magnitude of EXAFS oscillations in r-space for: (a) ZnO and ZnO composites with poly-PCDA in the blue and red phases; and (b) TiO$_2$ and TiO$_2$ composites with poly-PCDA in the blue and red phases.

## CONCLUSIONS

FTIR, Raman and EXAFS data for poly-PCDA nanocomposites with zinc oxide and titanium dioxide indicate bonding interactions only with zinc oxide involving the head group of the diacetylene side chain. This interaction results in the infrared activation of a C=C bond stretching mode at 1540 cm$^{-1}$, a sizeable decrease in intensity of the C=O stretching line in the FTIR spectrum, and the appearance of a new C≡C Raman line at 2259 cm$^{-1}$ in the red phase of the poly-PCDA-ZnO nanocomposite. This interaction probably leads to the reversibility of the red phase and sizeable increase of the chromatic transition temperature in the poly-PCDA-ZnO nanocomposite. Weak interactions between poly-PCDA blended with PVDF, cellulose and PVA, are also indicated by shifts of the Raman and FTIR spectra and increase in the blue to red transition temperatures, but these interactions are not strong enough to promote reversibility of the red phase in poly-PCDA blends with these polymers.

**ACKNOWLEDGEMENT:** This work has been supported by the US Army, ARDEC.

*Corresponding author: iqbal@adm.njit.edu

**REFERENCES:**
1. R. W. Carpick, D.Y. Sasaki, M.S.Marcus, M.A. Eriksson and A.R. Burns, *J.Physics: Condensed Matter* **16**, R679 (2004).
2. S. Lee and J-M. Kim, *J. Macromolecules* **40**, 26 (2007).
3. Y. Lu, Y. Yang, A. Sellinger, M. Lu, J. Huang, H. Fan, R. Hadded, G. Lopez, A.R. Burns, D.Y. Sasaki, J. Shelnutt and C.J. Brinker, *Nature* **410**, 913 (2001).
4. S. Dei, A. Matsumoto and A. Matsumoto, *Macromolecules* **41**, 2467 (2008).
5. Z. Iqbal, N.S. Murthy, Y.P. Khanna, J.S. Szobota, R.A. Dalterio and F.J. Owens, *J.Phys.C: Solid State Phys.***20**, 4283(1987).
6. M. Wenzel and G.H. Atkinson, *J.Amer.Chem. Soc.* **111**, 6123 (1989).
7. H. Eckhardt, D.S. Boudreaux and R.R. Chance, *J. Chem. Phys* **85,** 4116 (1986).
8. Y. Gu, W. Cao, L. Zhu, D. Chen, and M. Jiang, *Macromolecules* **41**, 2299 (2008).
9. D.J. Ahn, E-H. Chae, G.S. Lee, H-Y. Shim, T-E. Chang, K-D. Ahn, and J-M. Kim, *J. Amer.Chem. Soc.***125,** 8976 (2003).

Mater. Res. Soc. Symp. Proc. Vol. 1190 © 2009 Materials Research Society          1190-NN11-29

# Fabrication of Ionic Polymer Metal Composite Actuator With Palladium Electrodes and Evaluation of Its Bending Response

Takuma Kobayashi, Takeshi Kuribayashi and Masaki Omiya

Department of Mechanical Engineering, Keio University,
3-14-1, Hiyoshi, Kohoku-ku, Yokohama, Kanagawa, 223-8522, Japan

## ABSTRACT

We built up the way of fabricating ionic polymer metal composite (IPMC) actuator with palladium electrodes and evaluated the bending response under various solvents. We fabricated IPMC consisting of a thin Nafion® membrane, which is the film with fluorocarbon back-bones and mobile cations, sandwiched between two thin palladium plates. The surface resistivity was $2.88\pm0.18\Omega$/sq. , so it could be said to have enough conductive property. Then, we observed its cross section by using FE-SEM. As a result, palladium plates were evenly coated and its thickness was about $30\mu m$. Then, we evaluated the bending response under the various solvents. In the evaluation, we study the influence to bending response by cation forms and mol concentration. We used $Li^+$, $Na^+$ and $NH_4^+$ as cation form and all of them were 1.0mol/l. In terms of ionic radius, $Li^+< Na^+< NH_4^+$. As mol concentration, we use 0.1mol/l, 0.5mol/l, 1.0 mol/l and all of them were $Na^+$ form. As a result the bigger the ionic radius become, the larger bending response IPMC actuator showed, and the higher the ionic concentration become, the larger bending response IPMC actuator showed. Then, we conducted the modeling of IPMC actuator.

## INTRODUCTION

An ionic polymer-metal composite (IPMC) consisting of a thin perfuorinated ionomer membrane, electrodes plated on both faces, undergoes large bending motion when a small electric field is applied across its thickness in a hydrated state[1]. The characteristics of IPMC are ease of miniaturization, low density, and mechanical flexibility. Therefore, it is considered to have a wide range of applications from MEMS sensor to artificial muscle. However, there are problems on IPMC. First, it is high-priced because most of IPMC actuators use gold or platinum as electrodes [2, 3]. Second, its mechanical and electric characteristics under various solvents have not been clarified because of the complex mechanism of the deformation [4, 5]. In order for IPMC actuator to be widely put to practical use, we should solve these problems. Hence, this research focuses on fabrication of IPMC actuator with palladium electrode, which is cheaper than gold or platinum, and evaluation of its bending response under various solvents.

## EXPERIMENTAL DETAILS

### Fabrication of IPMC actuator with palladium electrodes

We used Nafion®117 ,whose thickness is 183μm, as thin perfuorinated ionomer membrane. Nafion® is chemically and thermally stable and excels in the electrochemical property[6]. It has been often used in the IPMC study. As electrodes, we chose palladium. Palladium is inferior in corrosion resistance and electric conductivity to gold and platinum, but it is quarter as expensive as they are, and one third as weighty as they are, so when we use palladium as electrodes, we can get cheaper and lighter IPMC.

We fabricated IPMC actuator with palladium electrodes by nonelectrolytic plating method. First, we pretreated Nafion® membrane. In the pretreatment, we saturated Nafion® membrane in the acetone (0.789 g/ml) for twenty seconds, and then we did sensitizer-activator process for Nafion® membrane. Second, we prepared the plating bath. The component of the plating bath is thiodiglycollic acid (0.003g), palladium chloride (0.18g) and ethylendiamine (1.00ml), sodium phospinate monohydrate (0.56g) for 100ml water. Then, we saturated Nafion® membrane in the plating bath for about 30 minutes. At this time, we kept the plating bath temperature about 328K. In this way, we got IPMC with palladium electrodes. Then, we observed the surface of electrodes and its cross section of IPMC by using FE-SEM, and measured the surface resistivity, dielectric constant, fractional increments of volume due to the water absorption, Young's modulus, yield stress, and tensile strength.

### Evaluation of Bending Response

Then, we evaluated the bending response of IPMC actuator with palladium electrodes under the various solvents. In the evaluation, we study the influence to bending response by cation forms and mol concentration. In the experiment about the influence by cation forms, we used $Li^+$, $Na^+$ and $NH_4^+$ as cation forms and all of them were 1.0mol/l. In terms of ionic radius, $Li^+<$ $Na^+< NH_4^+$. In the experiment about the influence by mol concentration, we used 0.1mol/l, 0.5mol/l, 1.0 mol/l as mol concentration and all of them were $Na^+$ form. In each solvent, we applied constant voltage (see figure 1) and step voltage (see figure 2) to IPMC actuator, and then measured its horizontal tip displacement. Figure 3 shows the experimental system for measuring horizontal tip displacement.

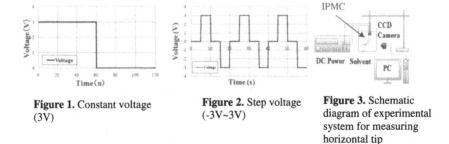

**Figure 1.** Constant voltage (3V)

**Figure 2.** Step voltage (-3V~3V)

**Figure 3.** Schematic diagram of experimental system for measuring horizontal tip displacement

# RESULTS

## Fabrication of IPMC actuator with palladium electrodes

Figure 4 shows IPMC we fabricated (4cm×1cm). The surface resistivity of electrodes was 2.88±0.18Ω/sq. , so it could be said to have enough conductive property. By observing the surface of electrodes (see figure 5) and its cross section of IPMC (see figure 6), it was proved that there were grains on the surface of IPMC, and that palladium plates were evenly coated and its thickness was about 30μm. Table I shows dielectric constant and fractional increments of volume of IPMC in $Li^+$, $Na^+$ and $NH_4^+$ form. It is proved that the larger the ionic radius become, the smaller dielectric constant and fractional increments of volume become. From table II, it is shown that Young's modulus and tensile strength tend to be smaller when cation radius of IPMC is larger.

 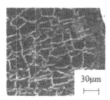 

**Figure 4.** IPMC with palladium electrodes

**Figure 5.** SEM photo of the surface of electrode

**Figure 6.** SEM photo of cross section of IPMC

**Table I.** Ionic radius and dielectric constant and fractional increments of volume under several cation forms

| Cation form | Ionic radius (Å) | dielectric constant (F/m) | fractional increments of volume(%) |
|---|---|---|---|
| $Li^+$ | 0.90 | 8.42 | 58.3 |
| $Na^+$ | 1.16 | 7.50 | 53.3 |
| $NH_4^+$ | 1.43 | 5.80 | 50.4 |

**Table II.** Young's modulus, yield stress and tensile strength of IPMC under several cation forms

| Cation form | Young's Modulus(MPa) | Yield stress (MPa) | Tensile stress (MPa) |
|---|---|---|---|
| dry | 286 | 17.5 | 20.4 |
| Li+ | 125 | 4.8 | 8.1 |
| Na+ | 104 | 5.1 | 7.7 |
| NH4+ | 92 | 5.0 | 5.2 |

## Evaluation of Bending Response

Figure 7 shows time-variation of tip displacement and current for constant voltage (Na⁺, 1.0mol/l). From this figure, it is proved that horizontal tip displacement increased with time of applied voltage. Figure 8 shows time-variation of tip displacement for constant voltage (3V) in Li⁺, Na⁺ and NH₄⁺ form (1.0mol/l). This figure showed that the bigger the ionic radius became, the larger bending IPMC

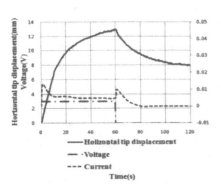

**Figure 7.** Time-variation of tip displacement and current for constant voltage (Na⁺, 1.0mol/l)

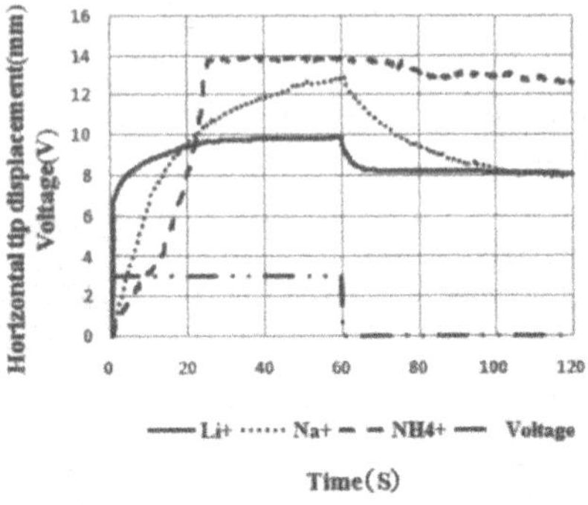

**Figure 8.** Time-variation of tip displacement for constant voltage (3V) in Li⁺, Na⁺ and NH₄⁺ form (1.0mol/l)

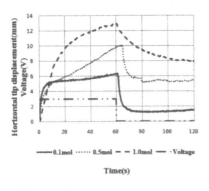

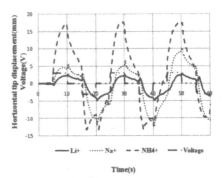

**Figure 9.** Time-variation of tip displacement for constant voltage (3V) in Na⁺ form (0.1, 0.5, 1.0mol/l)

**Figure 10.** Time-variation of tip displacement for step voltage (-3V~3V) in Li⁺, Na⁺ and NH₄⁺ form (1.0mol/l)

184

actuator showed. It is thought that this is because the bigger ion can cause large bias of volume between two sides. Figure 9 shows time-variation of tip displacement for constant voltage (3V) in Na$^+$ form (0.1, 0.5, 1.0mol/l).The higher the ionic concentration became, the larger bending IPMC actuator showed. Figure 10 shows time-variation of tip displacement for step voltage (-3V~3V) in Li$^+$, Na$^+$ and NH$_4^+$ form (1.0mol/l). In each cation form, IPMC actuator underwent good bending response. Especially, in NH$_4^+$ form, large bending was realized. Figure 11 shows time-variation of tip displacement for step voltage (-3V~3V) in  Na$^+$ form (0.1, 0.5, 1.0mol/l). In each mol concentration, IPMC actuator showed good bending response. Especially, in 1.0mol/l , large bending was showed.

Figure 12 shows one dimensional model of IPMC actuator. If we assume that the curvature of IPMC is constant when it is bending, the model we should consider is Figure 12, where M is moment applied at the tip of cantilever. The moment $M$ can be expressed as Equation 1 [3].

$$M \approx k_0 \kappa_e \phi_0 ahb \tag{1}$$

Where $k_0$ is cluster constant, $\kappa_e$ is the effective dielectric constant, $\phi_0$ is the applied potential, $a$ is constant, $h$ is half of thickness of IPMC, and $b$ is width of IPMC. At this time, the tip displacement $w$ of IPMC can be expressed as Equation 2.

$$w = \frac{M}{2E_c I_e} l^2 \tag{2}$$

Where $E_c$ is Young's modulus of electrodes, $I_e$ is the equivalent geometrical moment of inertia, and $l$ is the length of IPMC. In this way, we can obtain the tip displacement of IPMC. Figure 13 shows Comparison of experimental and theoretical value (cluster value = 86.5).

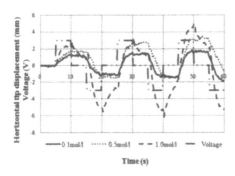

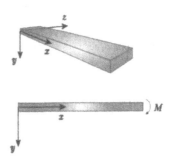

**Figure11.** Time-variation of tip displacement for step voltage (-3V~3V) in  Na$^+$ form (0.1, 0.5, 1.0mol/l)

**Figure 12.** One-dimensional model

185

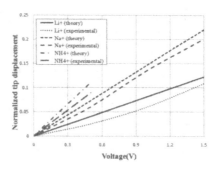

**Figure 13.** Comparison of experimental
and theoretical value
(cluster constant = 86.5)

## CONCLUSIONS

In this study, we built up the way of fabricating IPMC actuator with palladium electrodes, and evaluate its bending response in the various solvents. As a result, it was proved that the bigger the ionic radius became or the higher the ionic concentration became, the larger bending response IPMC showed. Then, we conducted modeling of IPMC, and compared experimental value to theoretical value.

## REFERENCES

1. Yoseph Bar-Cohen, Stewart Sherrit and Shyh-Shiuh Lih "Characterization of the Electromechanical Properties of EAP materials", Proceedings of EAPAD, SPIE's 8th Annual International Symposium on Smart Structures and Materials, Newport, CA. Paper No. 4329-43 (2001)
2. Sia Nemat-Nesser and Yongxian Wu,"Comparative experimental study of ionic polymer–metal composites with different backbone ionomers and in various cation forms", Journal of Applied Physics,**93**,5255 (2003)
3. S. Nemat-Nasser and J. Yu Li "Electrochemical response of ionic polymer-metal composites",
Journal of Applied Physics, 87(2000)
4. K.J. Kim, W. Yim, J.W. Paquette,and D. Kim, " Ionic Polymer-metal Composites for Underwater Operation", Journal of Intelligent Materials System and Structures (JIMSS), (2006, in print)
5. J.W. Paquette, K.J. Kim, "Ionmeric Electro-Active Polymer Artificial Muscle for Naval Applications", IEEE Journal of Oceanic Engineering (JOE), Vol.29, No.3, pp.729-737 (2004)
6. M.D. Bennett and D.J.. Leo, "Ionic Liquids as Solvents for Ionic Polymer Transducers, Sensors and Actuators A": Physical, Vol. 115. pp. 79-90 (2004)

Mater. Res. Soc. Symp. Proc. Vol. 1190 © 2009 Materials Research Society          1190-NN11-35

# Sensing Materials for the Detection of Chlorine Gas in Embedded Piezoresistive Microcantilever Sensors

Timothy L. Porter[1], Tim Vail[2], Amanda Wooley[2] and Richard Venedam[3]

[1]Northern Arizona University, Dept. of Physics, Flagstaff, AZ 86011, USA
[2]Northern Arizona University, Dept. of Chemistry, Flagstaff, AZ 86011, USA
[3]National Security Technologies, LLC, Las Vegas, NV 89193, USA

## ABSTRACT

Embedded piezoresistive microcantilever (EPM) sensors provide a small, simple and robust platform for the detection of many different types of analytes [1-4]. These inexpensive sensors may be deployed in battery-powered handheld units, or interfaced to small, battery-powered radio transmitter-receivers (motes), for deployment in mesh networks of many sensors. Previously, we have demonstrated the use of EPM sensors in the detection of hydrogen fluoride gas [5], organophosphate nerve agents [6], volatile organic compounds (VOC's) [2-4], chlorinated hydrocarbons in water, certain animals [7] and others [8-10]. Here, we report on the design of EPM sensors functionalized for the detection of chlorine gas, or $Cl_2$. We have constructed EPM sensors using composite materials consisting of a polymer or hydrogel matrix loaded with agents specific for the detection of $Cl_2$ such as NaI. These materials were tested in both controlled laboratory conditions and in outdoor releases. Results are presented for gas exposures ranging from 1000 ppm to 20 ppm.

## INTRODUCTION

When chlorine gas reacts with water, such as what might occur when inhaled, both hydrochloric acid (HCL) and hypochlorous acid (HOCL) are formed. Upon exposure in mammals, inflammation of the conjunctivae, nose, pharynx, larynx, trachea, and bronchi may occur. In animal studies of chlorine gas exposure, immediate respiratory arrest occurs at 2000 ppm, with the lethal concentration for a 50% mortality rate of exposed animals occurring in the range of 800-1000 ppm. For exposure in humans, symptoms that may occur include coughing, chest tightness, burning in the chest and lungs, blurred vision, nausea, vomiting, difficulty breathing and pulmonary edema.

Sensors for the detection of chlorine gas may be important from both an industrial standpoint and for military or civil defense use. Small, portable sensors using piezoresistive microcantilevers (PMC) are well suited for such needs [3]. A photograph of a single PMC die is shown in Fig. 1. These sensors may be deployed as battery-powered handheld sensors, or deployed in larger numbers in a mesh network of "motes" interfaced through tiny radio frequency units [11]. Each of these PMC sensors utilizes a tiny piezoresistive microcantilever in contact with or partially embedded into a sensing material. The sensing material is synthesized in such a way as to react volumetrically upon exposure to the particular analyte to be detected.

Some sensing materials may be common polymers, which may react to a range of different analytes [3]. In this case, small sensor arrays may be fabricated utilizing many different materials to produce a unique signature for each analyte to be detected. In other cases, the sensing material is designed to respond to only a single analyte, as may be the case with some types of biological sensors [12]. In either case, when the sensing material responds to the presence of the analyte, the volumetric change in the sensing material causes the microcantilever to bend slightly, which may be read-out as a simple resistance change in the cantilever.

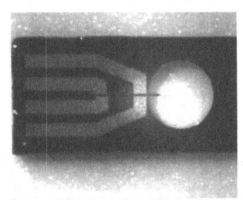

Fig. 1. Micrograph of piezoresistive microcantilever die. Each die contains a microcantilever extending partially into a circular area, and an integrated thermistor for temperature correction when needed. The dimensions of the cantilever are approximately 200 microns in length, 40 microns in width, and 1 micron thick. The nominal resistance of these cantilevers is 2000 $\Omega$. The sensitivity to deflection is approximately 4-5 $\Omega$ per micron of deflection.

## EXPERIMENTAL

Two different sensing materials were used in this study. In each case, a host material was used to contain sodium iodide crystals (NaI). The first host material was Hypol, a water-dispersible polyurethane material manufactured by Dow Chemical. The second host material was Dow Corning Sylgard 184, a poly(dimethylsiloxane) compound which will cure at lowered temperatures in 24 – 48 hours. In each case, the NaI crystals were mixed into the host polymer 30% by weight. While still in liquid form, individual sensors were fabricated by inserting the microcantilever tip (30 – 40 microns) into the sensing material. The assembly was then cemented into place and the sensing material was allowed to cure. After a 24-hour drying period, the sensors were ready for use. Sensor responses noted below were not affected by longer drying/curing times.

Laboratory experiments were performed by interfacing the sensors to a laptop computer directly through the USB interface [1]. Tests conducted in an outdoor environment (Nevada Test Site) utilized radio-frequency motes interfaced to the sensors. These motes have a range of about 400 m, and their data is received by a laptop computer interfaced to a mote base station. In Fig. 2 below, photos of the USB laptop interface and a single sensor mote are shown.

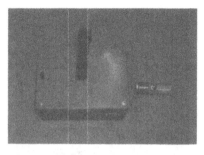

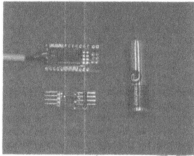

Fig. 2. (top) Single battery-powered mote interfaced to Cl sensor. The mote is powered by AA batteries, and has a range of approximately 400m. (bottom) Photo of USB sensor interface. Two chips are used for this interface, a U421 board (top) and AD7793 (bottom).

Chlorine exposures in the lab took place in a vented chamber of approximately 3.5 L volume. $Cl_2$ gas was added in controlled amounts so that a precise determination of exposure levels could be calculated. Outdoor exposures took place at the DOE Nevada Test Site Non-Proliferation Test and Evaluation Center (NPTec). Chlorine gas was released through portable stacks from gas bottles in a controlled release lasting for 15-30 min. The EPM sensors interfaced to radio motes were positioned 80 m downwind from the release points. The exact exposure levels were not known, but they were estimated by comparison from the controlled laboratory experiments.

**RESULTS AND DISCUSSION**

In Fig. 3 below, we show a 180 micron SEM image of sensing material composed of poly(dimethylsiloxane) and NaI Crystals. The crystals are highly visible in this image as the lighter contrast components of this material. After sensor assembly, this material (and the Hypol-based material) both exhibited good material properties with respect to ductility, stability and durability. Sensors fabricated from both materials operate over a year after initial assembly.

Fig. 3. 180 micron SEM image of sensing material composed of poly(dimethylsiloxane) and NaI Crystals. The crystals are highly visible in this image as the lighter contrast components of this material.

In Fig. 4, we show the sensor responses to controlled $Cl_2$ exposure at 1000 ppm. Initial sensor exposure took place at time 125 s, with both sensors responding initially within 5 sec of exposure. The plot with the larger response (series 1) is for the sensing material containing the poly(dimethylsiloxane) matrix, while the lower response corresponds to the sensor with the Hypol matrix. The sensors continued to respond well after 15 min. of total exposure (not shown) before finally becoming saturated. Fig. 5 shows a sensor fabricated using the poly(dimethylsiloxane) matrix with no NaI added. The response here is essentially zero for this "blank" sensor.

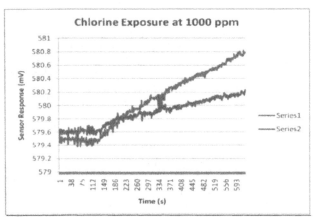

Fig. 4. Sensor responses to $Cl_2$ at 1000 ppm in the laboratory. The plot with the larger response (series 1) is for the sensing material containing the poly(dimethylsiloxane) matrix, while the lower response corresponds to the sensor with the Hypol matrix. In both cases, the initial exposure to $Cl_2$ occurred at time 125s, with the initial response from both sensors in less than 5 s.

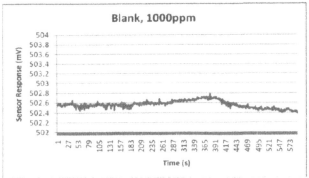

Fig. 5. Exposure of poly(dimethylsiloxane) sensor with no NaI component to 1000 ppm Cl$_2$.

Table 1 summarizes the responses from the sensors at lower exposures to Cl$_2$. At the 100 ppm exposure level, the poly(dimethylsiloxane)-based sensor again exhibits the larger of the two responses, with the Hypol sensor at approximately half the total response over 300 s.

| Exposure Level | 100 ppm, poly(dimethylsiloxane) | 100 ppm Hypol | Unknown Outdoor poly(dimethylsiloxane) |
|---|---|---|---|
| Sensor Response (mV) | 0.2 | 0.1 | 0.035 |

Table 1. Sensor responses at low levels of Cl$_2$ exposure.

We also exposed a poly(dimethylsiloxane)-based sensor to an unknown Cl$_2$ dose in an outdoor environment. Here, Cl$_2$ was released from a tank 80m upwind from the sensor location. The total sensor response after 300 s indicated that the Cl$_2$ vapor had a concentration of approximately 17-20 ppm at the sensor location. Finally, these sensors act as integrating devices, accumulating total response non-reversibly over long periods of Cl$_2$ exposure. For very light initial exposures, sensors generally react as new sensors, however very large initial exposures can render the sensors unresponsive, acting as single use chemical fuses.

## CONCLUSIONS

We have demonstrated working Cl$_2$ sensors using piezoresistive microcantilever technology. Sensing materials used a polymeric matrix loaded with NaI crystals, with poly(dimethylsiloxane)-based sensors consistently producing larger responses (70 – 100%) that sensors based on Hypol hydrogel. Poly(dimethylsiloxane)-based sensors exhibited sensitivities to detection of Cl$_2$ down to approximately 20 ppm in an outdoor environment. Future work on these sensors will include utilizing other polymer or gel matrices, smaller, more uniform NaI particles, and a wider range of loading percentages of NaI within the host material.

## ACKNOWLEDGEMENTS

We would like to thank the staff at the DOE Nevada Test Site NPTec facility for their excellent level of support and cooperation in these and other tests.

## REFERENCES

[1]     T. L. Porter and W. Delinger, "LabView based piezoresistive microcantilever sensor system," *Sensors and Transducers,* vol. 68, pp. 568-574, 2006.

[2]     T. L. Porter, M. P. Eastman, C. Macomber, W. G. Delinger, and R. Zhine, "An embedded polymer piezoresistive microcantilever sensor," *Ultramicroscopy,* vol. 97, pp. 365-369, 2003.

[3]     T. L. Porter, M. P. Eastman, D. L. Pace, and M. Bradley, "Polymer based materials to be used as the active element in microsensors: a scanning force microscopy study," *Scanning,* vol. 22, pp. 1-5, 2000.

[4]     T. L. Porter, M. P. Eastman, D. L. Pace, and M. Bradley, "Sensor based on piezoresistive microcantilever technology," *Sensors and Actuators,* vol. A88, pp. 47-51, 2001.

[5]     T. L. Porter, T. Vail, J. Reed, and R. Stewart, "Detection of Hydrogen Fluoride Gas Using Piezoresistive Microcantilever Sensors," *Sensors and Materials,* vol. 20(2), pp. 103-110, 2008.

[6]     T. L. Porter, T. Vail, and R. Venedam, "Detection of Organophosphate Gases and Biological Molecules using Embedded Piezoresistive Microcantilever Sensors," *Proc. Mater. Res. Soc.,* vol. 1086E, pp. 1086-u08-02, 2008.

[7]     T. L. Porter, T. R. Dillingham, and R. J. Venedam, "A microcantilever sensor array for the detection and inventory of desert tortoises," *Applied Herpetology,* vol. 5(3), pp. 293-301, 2008.

[8]     R. L. Gunter, W. Delinger, T. L. Porter, R. Stewart, and J. Reed, "Hydration level monitoring using embedded piezoresistive microcantilever sensors," *Medical Engineering and Physics,* vol. 27, pp. 215-220, 2005.

[9]     R. L. Gunter, W. G. Delinger, K. Manygoats, A. Kooser, and T. L. Porter, "Viral detection using an embedded piezoresistive microcantilever sensor," *Sensors and Actuators (A),* vol. A107, pp. 219-224, 2003.

[10]    A. Kooser, R. L. Gunter, W. G. Delinger, T. L. Porter, and M. P. Eastman, "Gas sensing using embedded pezoresistive microcantilever sensors," *Sensors and Actuators,* vol. 99(2-3), pp. 430-433, 2004.

[11]    T. L. Porter, W. Delinger, and R. Venedam, "Gas Detection Using Embedded Piezoresistive Microcantilever Sensors in a Wireless Network," *Sensors and Transducers,* vol. 94(7), pp. 133-138, 2008.

[12]    T. L. Porter, W. Delinger, A. Kooser, K. Manygoats, and R. Gunter, "Viral Detection Using an Embedded Piezoresistive Microcantilever Sensor," *Sensors and Actuators (A),* vol. 107(3), pp. 219-224, 2003.

Mater. Res. Soc. Symp. Proc. Vol. 1190 © 2009 Materials Research Society          1190-NN11-38

## Tests of the Affinity Assumption in Phantomlike Elastomer Networks

Misty Davies and Adrian Lew
Mechanical Engineering, Stanford University, Stanford, CA 94305, U.S.A.

ABSTRACT

Phantomlike elastomer simulations do not always deform globally affinely in the way that classical theory predicts. Assuming that each crosslink will deform affinely with its topological neighbors gives much better results, and creates a way to isolate crosslinks with unpredictable deformation properties. The correlation of non-affinities and network properties depends on the constitutive model and boundary condition used. We always find a correlation between local density of crosslinks and degree of non-affinity.

INTRODUCTION

The assumption that elastomers will deform affinely everywhere if the boundary deforms affinely ("global affinity") is often made [11, 9]. Recent experiments [4, 23, 22, 21, 28] have shown that these networks do not deform globally affinely at scales below a chain length. This difference between the deformation at the largest scales and at the smallest is considered by J.E. Mark [18] to be "one of the central problems in rubberlike elasticity". Non-affinity for some materials can be partially explained by variations in chain stiffness [12]. Other proposed causes include crosslink density [16, 3, 4, 23, 7, 21, 25, 26, 17, 6, 27, 10, 6], chain reorientation [20, 8], finite extensibility [22,2,1], strain-induced crystallization [14,2,1], and interchain excluded volume terms [25, 19, 26, 27]. It is likely is that there are many different simultaneous causes.

A hypothesis proposed here is that these networks will deform *topologically locally affinely* ("locally affinely"). This means connected crosslinks deform affinely together, while crosslinks that are in the near neighborhood and unconnected may not. Given a crosslink in an elastomer network, there will be many more unconnected crosslinks than connected in the geometrically local vicinity. Affinity and network property correlations of affinity are studied here for models with and without intrachain excluded volume terms. We also test models with and without finite extensibility. All models tested here neglect interchain effects.

THEORY

We use simulations with five different energy models and two different boundary conditions (10 total cases) focused at the scale of the individual crosslink. Our network is randomly generated with 1000 crosslinks and 1998 chains within a unit cube—crosslinks have functionality 3 or 4. The simulation method is incapable of capturing interchain interactions. The *Fixed* boundary condition constrains all of the crosslinks within a small distance from the outside edge; the *Free* boundary condition constrains crosslinks at the bottom and top of the cube, but leaves the crosslinks along the sides free and does not preserve incompressibility—this creates a network that looks like it is experiencing necking. (See Figure 1.)

For each of the constitutive model equations below, $A$ is the Helmholtz free energy and $r$ is the length of an individual chain. The benchmark model is the Entropic Spring relationship, as given by Equation 1. For the *Fixed* condition and this model, interior crosslinks deform perfectly

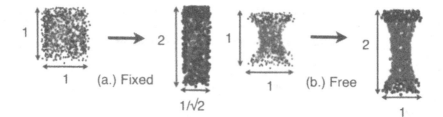

**Figure 1: The boundary conditions used to test affinity within this paper. Darker crosslinks are fixed and lighter crosslinks are used to minimize the energy.**

globally affinely. The second model (Equation 2) is the Langevin Spring [13]. We minimize computational cost by finding $\beta$ using a series approximation to the Langevin equation. The Langevin Spring model captures finite extensibility effects, but—like Entropic Spring—the equilibrium chain length is zero. Our Real Chain model (Equation 3) uses a self-avoiding walk length distribution as shown in Rubinstein and Colby [24] in Section 3.2 to generate finite-length chains at equilibrium. The final two models define the reference as the zero-energy state and measure the energy change as a departure. The first such model is the Self-Avoiding Gaussian (Equation 4) [12, 16, 8, 11]. These chains have intrachain excluded volume interactions; the generated network is the zero-energy state and is not zero-volume. The Excluded Volume Quadratic relationship (Equation 5) also captures some strain-stiffening. The five equations are:

$$A(r)_{ES} = \frac{1.5kT}{b^2 g} r^2 \tag{1}$$

$$A(r)_{LS} = r\beta(r) + \ln \frac{\beta(r)}{\sinh \beta(r)} \tag{2}$$

$$A(r)_{RC} = kT \left( \frac{r}{bg^\nu} \right)^{\frac{1}{1-\nu}} \tag{3}$$

$$A(r)_{SG} = \frac{\mu r_0}{2} \left( \frac{r - r_0}{r_0} \right)^2, \text{ and} \tag{4}$$

$$A(r)_{EVQ} = \mu \left( \frac{r^2 - r_0^2}{r_0^2} \right)^2, \tag{5}$$

where $k$ is the Boltzmann constant, $T$ is the temperature (assumed to be constant), $g$ is the number of Kuhn lengths in the chain [15], $b$ is length of a Kuhn segment, $r_0$ is the original length of the chain and $\mu$ is a constant that contains thermal and chain stiffness information. We've chosen to use Flory's estimate of the fractal dimension of a real chain, so that $\nu = 0.588$.

A crosslink deforms affinely if its new vector position $\bar{x}$ can be given by

$$\bar{x} = \mathbf{F}(\bar{X} - \bar{X}_0) + \bar{\Delta} + \bar{X}_0 \tag{6}$$

where $\mathbf{F}$ is a second-order tensor, $\bar{X}$ is the reference position of the crosslink, $\bar{X}_0$ is the reference condition of some point in space, and $\bar{\Delta}$ is the translation of $\bar{X}_0$. Each test case compares the positions of the crosslinks in the reference position against the positions of the crosslinks in the deformed configuration. The reference positions are found by minimizing the energy [5] of the generated networks without deformation. The new configurations are found by minimizing the network energy after deformation using the free crosslink positions.

## DISCUSSION

### Global Affinity Within Elastomer Networks

We define a global affinity error $\xi_{global}$ for each crosslink that is equal to the difference in position between the crosslink in its deformed state $\bar{x}$ and the position the crosslink would have had if it had deformed from its reference position $\bar{X}$ with the same affine deformation tensor $\mathbf{F}$ used to deform the boundary nodes. The distance is normalized by the average undeformed length of a chain in the network; each error is then in percent of chain lengths. The formula is

$$\xi_{global} = \frac{\left\| \bar{x} - \mathbf{F}\bar{X} \right\|}{\langle \bar{R} \rangle} \tag{7}$$

where $\langle \bar{R} \rangle$ is the average reference length of the chains. The global network affinity $\Xi_{global}$ is then just the mean crosslink affinity error $\xi_{global}$ for each case.

The global affinity assumption only applies when the boundaries deform affinely; we focus here on the *Fixed* boundary condition only. Table 1 shows the network percent $\Xi_{global}$ each of the simulations is from globally affine. The Entropic Spring model is the only model without non-affinities. For Langevin, the mean deformed length was 21% longer than the reference length, and the maximum deformed length was 77%. The longest chains did not have the strongest correlation to non-affinity. We found that finite extensibility and intrachain excluded volume terms were the strongest contributors. The highly non-affine crosslinks are distributed spatially throughout the network. These trends are true both for cases that include finite extensibility but neglect intrachain excluded volume, and vice versa.

### Local Affinity Within Elastomer Networks

To explore local affinity, we examined both the previous test cases and the *Free* boundary condition cases. We first find the original average position of a connected set of crosslinks and use this position as $\bar{X}_0$, then find the translation $\bar{\Delta}$ of $\bar{X}_0$ by looking at its new position. We define two new variables for each crosslink: $\bar{Y} = \bar{X} - \bar{X}_0$ and $\bar{y} = \bar{x} - \bar{\Delta} - \bar{X}_0$. We find the components of $\mathbf{F}_{local}$ for each connected crosslink $i$ by minimizing

$$\sum_i \left\| \bar{y}_i - \mathbf{F}_{local}\bar{Y}_i \right\|^2. \tag{8}$$

Once the local affine deformation matrix $\mathbf{F}_{local}$ is defined for each crosslink, the error in local affinity $\xi_{local}$ is defined as

$$\xi_{local} = \frac{\left\| \bar{x} - \mathbf{F}_{local}\left(\bar{X} - \bar{X}_0\right) - \bar{\Delta} - \bar{X}_0 \right\|}{\langle \bar{R} \rangle}, \tag{9}$$

and the network error $\Xi_{local}$ is the mean of all of the crosslink $\xi_{local}$.

An examination of Table 1 shows that the average network errors for the locally affine cases run between 2 percent and 12 percent of a chain length (compared to 6 to 74 percent in the global case). The *Free* boundary condition cases were not expected to deform affinely. Image results (not included here due to space constraints) show that there are no crosslinks more than one chain length from their locally affine positions and many crosslinks are more than one chain from their globally affine positions. The non-deforming crosslinks are scattered throughout the volumes.

An examination of the global versus local cases reveals that any method that assumes the entire network will deform affinely—even a method with energy models that neglect intrachain excluded volume interactions—will have large errors at the crosslink deformation level. By

| Constitutive Model | Boundary Condition | $\Xi_{global}$ | $\Xi_{local}$ |
|---|---|---|---|
| Entropic Spring | Fixed | 0% | 0% |
| | Free | 19% | 0% |
| Langevin Spring | Fixed | 7% | 3% |
| | Free | 30% | 2% |
| Real Chain | Fixed | 6% | 3% |
| | Free | 25% | 5% |
| Self-Avoiding Gaussian | Fixed | 14% | 8% |
| | Free | 74% | 12% |
| Excluded Volume Quadratic | Fixed | 16% | 8% |
| | Free | 66% | 10% |

Table I : Error in the network for both global and local affinity for two different boundary conditions and five different constitutive models. In the Boundary Condition column *Fixed* refers to the boundary condition in which all sides are fixed, while *Free* refers to the boundary condition in which only the top and bottom are fixed while the sides are left free.

contrast, assuming that a crosslink will deform affinely with its topological neighbors results in much smaller errors. One limitation of this study is that we have made no attempt to find large groups of crosslinks that are deforming affinely together, since the local affinity error depends only on nearest neighbors. Another limitation is that we also make no attempt to discover whether the crosslinks that are not deforming the least locally affinely are topological neighbors.

### Correlations of Affinity With Various Elastomer Network Properties

The most likely non-affinity causing properties were the original orientation and length of the chains, and the original local density of the crosslinks. In this work we define the local crosslink density as the number of crosslinks within one average chain length. A chain is determined to be non-affine if the local affine deformation matrices $F_{local}$ for its two crosslinks are significantly different. The metric $\tau$ is

$$\tau = trace\left(A^T A\right) . \tag{10}$$

where $A$ is the difference between the two $F_{local}$ matrices. A chain is flagged as a non-affine chain if its $\tau$ value is more than one standard deviation above the mean of $\tau$.

We created a histogram for each property and keep track of both the total number of chains and the number of non-affine chains that go in each bin. For the original length of the chains this is straightforward. For the original orientation of the chains, we take the dot product of the reference chain orientation with the primary stretching axis as a measure of how aligned each chain is with that axis. For the local density measurement, we define each chain's density as the average of the crosslink densities for the two crosslinks at its endpoints. The density of each crosslink is the number of other crosslinks within one average chain length. After we have created each histogram, we create a line graph plotting the percentage of chains in each bin that are non-affine. Unfortunately, the images cannot be shown here due to space constraints. The total number of chains and the percentage of chains that are non-affine have been calculated for each bin. Note that there may be very few chains within a bin on the histogram, and that this affects the interpretation of the percentage.

None of the cases tested here demonstrate the strong correlation between non-affinity and alignment with the stretching axis suggested by Chandran, et al. [9]. If anything, there may be a slight tendency for Self-Avoiding Gaussian, Excluded Volume Quadratic or Langevin Spring chains oriented about 45 degrees from the stretching axis to deform non-affinely. However, since only 4 to 10 percent of the chains at about 45 degrees from the stretching axis are non-affine and the trend does not hold across all of the test energy models, we assume no relationship between direction and affinity for these phantomlike network simulations.

Another suggestion in the literature is that non-affinity occurs because of the finite extensibility of chains in the network–that chains that are longer will be more likely to deform non-affinely because the energy cost for stretching is too high, and other chains must then take up the task of deforming to carry the stress [23, 2, 1]. For the cases tested here, only the Langevin Spring energy model attempted to capture the finite extensibility of chains in the network. However, even for this case we don't see an increase in non-affinity with chain length. If anything, there is a trend across all of these cases for the shorter chain lengths to deform non-affinely. The histogram for reference crosslink density is skewed heavily toward the center, with only 1 to 11 chains in the bin for highest density. The next highest density bin has between 18 and 109 chains for all of the cases tested here. An examination of the results, keeping in mind that there are very few chains in the highest density bins, shows that density is correlated with non-affinity across all of the tests except for the Real Chain model with the Free boundary condition: 10 to 23 percent of the chains that have the highest local density metrics are non-affine.

The correlation of density with non-affinity in the literature is associated with local stiffness variations. That effect is not expected to be at work here since we are neglecting interchain excluded volume terms. One possible explanation for the density effect in this network model is that clusters of shorter chains in the more dense regions may be able to freely rotate to carry more of the load without having to stretch. Since the energy models depend only on the length of the chains, this means that a lower energy configuration can be achieved by rotating these shorter chains.

## CONCLUSIONS

The affinity assumption for elastomer networks holds only in simulations that use the Entropic Spring model with perfectly affine boundary deformations. Simulations with other energy models and/or boundary conditions should calibrate with exact in order to quantify the potential error. A coarse-grained method assuming topologically-connected parts of the network will deform affinely should have better accuracy than one based on geometric regions. Non-affinities tend to occur where the crosslinks are the densest. This result has been explained by stiffness due to monomer packing. However, these simulations do not include monomer-packing effects, and also show a correlation between local crosslink density and non-affinities within the network. Further experimentation needs to be done to determine if there is a cause for the density-affinity correlation that applies both to phantomlike network models and physical elastomer networks.

## REFERENCES

1. A.L. Andrady, M.A. Llorente, and J.E. Mark, *J. Chem. Phys.* **73**, 1439 (1980).
2. A.L. Andrady, M.A. Llorente, and J.E. Mark, *J. Chem. Phys.* **72**, 2282 (1980).
3. J. Bastide and L. Leibler, *Macromolecules* **21**, 2647 (1988).
4. J. Bastide, L. Leibler, and J. Prost, *Macromolecules* **23**, 1821 (1990).
5. S. J. Benson, L. C. McInnes, J. Moré, T. Munson, and J. Sarich, http://www.mcs.anl.gov/tao.
6. A. F. Bower and J. H. Weiner. *J. Chem. Phys.* **120**, 11948 (2004).
7. R. Bruinsma and Y. Rabin, *Phys. Rev. E: Stat. Phys., Plasmas, Fluids* **49**, 554 (1994).
8. P.L. Chandran and V.H. Barocas, *J. Biomech. Eng.* **128**, 259 (2006).
9. A.A. Darinskii, A. Zarembo, and N.K. Balabaev, *Macromol. Symp.* **252**, 101 (2007).
10. B.A. DiDonna and T.C. Lubensky, *Phys. Rev. E: Stat. Phys., Plasmas, Fluids* **72**, 06619 (2005).
11. B. Erman and J.E. Mark, *Structures and Properties of Rubberlike Networks.* (Oxford University Press, New York, 1997).
12. D.A. Head, A.J. Levine, and F.C. MacKintosh, *Phys. Rev. Lett.* **91**, 108102 (2003).
13. H.M. James and E. Guth, *J. Chem. Phys.*, **11**, 455 (1943).
14. M.A. Kennedy, A. J. Peacock, and L. Mandelkern, *Macromolecules* **27**, (1994).
15. W. Kuhn and F. Grun, *J. Polym. Sci.* **1**, 183 (1946).
16. A.J. Levine, D.A. Head, and F.C. MacKintosh, *J. Phys.: Condens. Matter* **16**, S2079 (2004).
17. D. Long and P. Sotta, in *The IMA Vol. in Mathematics and its Applications: Modeling of Soft Matter, Vol. 141*, edited by M-C.T.Calderer and E.M. Terentjev (Springer, New York, 2005), pp. 205-234.
18. J.E. Mark, *J. Phys. Chem. B* **28**, 1205 (2003).
19. G. Marrucci, F. Greco, and G. Ianniruberto, *J. Rheol.*. **44**, 845 (2000).
20. P.R. Onck, T. Koeman, T. van Dillen, and E. van der Giessen, *Phys. Rev. Lett*, **95**, 178102 (2005).
21. A. Ramzi, A. Hakiki, J. Bastide, and F. Boue, *Macromolecules* **30**, 2963 (1997).
22. A. Ramzi, F. Zielinski, J. Bastide, and F. Boue, *Macromolecules* **28**, 3570 (1995).
23. C. Rouf, J. Bastide, J. M. Pujol, F. Schosseler, and J. P. Munch, *Phys. Rev. Lett.* **73**, 830 (1994).
24. M. Rubinstein and R. Colby, *Polymer Physics* (Oxford University Press, ??, 2003).
25. M. Rubinstein and S. Panyukov, *Macromolecules* **30**, 8036 (1997).
26. J. Sommer and S. Lay, *Macromolecules* **35**, 9832 (2002).
27. C. Svaneborg, G.S. Grest, and R. Everaers, *Phys. Rev. Lett.* **93**, 257801 (2000).
28. S. Westermann, W. Pyckhout-Hintzen, D. Richter, E. Straube, S. Egelhaaf, and R. May, *Macromolecules* **34**, 2186 (2001).

Mater. Res. Soc. Symp. Proc. Vol. 1190 © 2009 Materials Research Society          1190-NN12-03

## Hybrid Polymer/Ultrathin Porous Nanocrystalline Silicon Membranes System for Flow-Through Chemical Vapor and Gas Detection

Maryna Kavalenka[1], David Fang[1], Christopher C. Striemer[1,3], James L. McGrath[2] and Philippe M. Fauchet[1,2]
[1]Electrical and Computer Engineering, University of Rochester, Rochester, NY, USA
[2]Biomedical Engineering, University of Rochester, Rochester, NY, USA
[3]SimPore, Inc., West Henrietta, NY, USA

### ABSTRACT

Here we discuss a novel capacitive-type chemical sensor structure that uses recently discovered porous nanocrystalline silicon (pnc-Si) membranes [1] covered with metal as the capacitor plates while a polymer layer sandwiched between them serves as the sensing layer for solvent vapor detection. Pnc-Si is new ultrathin (15 nm) membrane material with pore sizes ranging from 5 to 50 nm and porosities from < 0.1 to 15 % that is fabricated using standard silicon semiconductor processing techniques. We present a study of pnc-Si membranes as a platform for such a sensor. The degree of swelling and the reversibility of the polymer/pnc-Si membrane system exposed to analyte-containing vapors are observed using optical and electrical techniques.

### INTRODUCTION

Highly sensitive sensors capable of chemical vapor and gas detection are an increasing demand in industry, medicine and environmental monitoring applications. Polymer based sensors are attractive because of their low cost and the variety of different polymers available for sensing applications. The polymer materials are used as responsive coatings as they are able to selectively absorb different molecules, which results in polymer swelling or a change in its electrical properties.

The transducer elements for polymer-based sensors include chemiresistors that measure resistance changes [2], chemicapacitors that detect changes in dielectric properties or polymer layer thickness [3-6], and mechanical oscillators that respond to mass changes [7,8]. Capacitive-type polymer based sensors can measure the response of a polymer film as a function of gas or vapor concentration and are the most preferable type of electrical sensors because of good response times, low-cost, low-power and relative ease of fabrication [3-6]. In such sensors, the polymer is used as a selective dielectric layer in a capacitor whose capacitance increases (decreases) as the analyte is absorbed into the polymer to change its permittivity (thickness).

The capacitive sensors have two geometries. The first geometry is based on Interdigitated Electrodes (IDE) [4, 5]. The sensor consists of two comb-shaped metal electrodes deposited on a substrate and a polymer deposited on top of them. The second type is a parallel-plate sensor that consists of a polymer layer sandwiched between two electrodes. The top electrode in the parallel-plate geometry must be porous for the analyte to reach to the sensing polymer. Etching voids in a top metal layer is challenging because metal etchants would destroy the polymer deposited under metal layer. One solution it is to create a parallel-plate MEMs structure with an etched porous electrode on top of the silicon wafer and then infiltrate the polymer later. This however requires sophisticated fabrication steps [3, 9]. The IDE based sensors are easier to make but they lack sensitivity as most of the electric field lines pass through the substrate, while in parallel-plate sensor the lines pass through the polymer.

To overcome the existing difficulties in making parallel-plate geometry sensor structures [3, 9] we propose to use a nanoporous membrane material fabricated using standard silicon semiconductor processing techniques as an analyte-permeable electrode. The nanosmooth surface of the porous electrodes will also allow making ultrathin sensitive polymer layers to improve response and recovery times in the future.

Our sensor structure uses recently discovered porous nanocrystalline silicon (pnc-Si) membranes covered with metal to form capacitor plates while the polymer layer sandwiched between them serves as the sensing layer. Pnc-Si is new ultrathin (15 nm) membrane material [1] with pore sizes 5-50 nm and porosities up to 15 %. The pores in the material are formed spontaneously during rapid thermal annealing of an ultrathin a-Si layer sandwiched between two thin SiO$_2$ layers. These membranes have very good mechanical strength and flexibility, and are solvent and temperature stable. Here we present a study of pnc-Si membranes as a platform for capacitive sensing.

The polymer layer is spin-coated on the membrane and covers the pores. The pore openings then serve as nanometer diameter areas available for vapor adsorption. The degree of swelling and the reversibility of the polymer/pnc-Si membrane system immersed in analyte-containing vapors are observed using optical and electrical techniques. Preliminary results of the electrical response of the pnc-Si membrane based sensor to the polymer swelling and change of the dielectric constant in the presence of various analytes are discussed.

## EXPERIMENTAL DETAILS

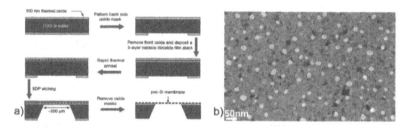

**Figure 1.** a) Pnc-Si membrane fabrication steps [1]; b) TEM image of a pnc-Si membrane where the white areas are the pores, the grey and black areas are silicon nanocrystals.

The pnc-Si membrane fabrication procedure is described in Fig.1a [1]. First, a thick SiO$_2$ layer thermally grown on both sides of a silicon wafer is patterned using standard photolithography to create a mask for membrane formation. Then the front oxide layer is removed and a three layer film stack (20 nm SiO$_2$/15 nm a-Si /20 nm SiO$_2$) is deposited on the front surface using RF magnetron sputtering. The structure is then treated at high temperature in a rapid thermal processing chamber. This treatment crystallizes the amorphous silicon film, forming a nanocrystalline film with voids that become the membrane pores. The patterned back side of the wafer is then etched with EDP (ethylenediamine pyrocatechol), which removes the silicon wafer along (111) crystal planes until it reaches the first SiO$_2$ layer of the film stack. Lastly, the three layer membrane is exposed to buffered oxide etchant (BOE) to remove the protective oxide layers, leaving only the freely suspended ultrathin pnc-Si membrane. The membranes fabricated this way can have different dimensions, but the tests described in this paper were done on the chips that have a 3 x 3 array of 100 μm x 100 μm membranes.

A transmission electron microscopy (TEM) image of a pnc-Si membrane is shown in Fig.1b. The nearly circular white areas are the open pores in the membrane and the grey is the nanocrystalline silicon. The darker areas in the image are nanocrystals with crystal plane alignment satisfying the Bragg condition, so that electrons are diffracted and these nanocrystals appear black in this bright-field image.

**Figure 2.** Hybrid pnc-Si membrane/PDMS/Si sensor fabrication steps.

The sensor fabrication process is illustrated in Fig.2. A 20 nm metal layer is first deposited on top of the pnc-Si membrane and on a silicon wafer chip using e-beam evaporation to create two electrode surfaces. This layer consists of a 10 nm of titanium adhesion layer under 10 nm of gold. The gold deposition on titanium is necessary because of the quick oxidation of the titanium layer. Titanium and gold are deposited in one run without breaking the vacuum. The next step is polymer deposition. In the preliminary experiments we chose to use polydimethylsioloxane (PDMS) Sylgard 184 from Dow Corning Corp. which is an easy to use and widely available polymer. The elastomer is first mixed with its curing agent in 10:1 ratio and then placed in a desiccator to remove the bubbles created while mixing. The PDMS layer is created by spin-coating pre-polymer on silicon at 6000 rpm for 3 min. The thickness of the resulted PDMS is approximately 10 μm based on the spin rate and duration. The pnc-Si chip and PDMS covered silicon chip are then bonded together and cured at 95°C for 2 hours.

## DISCUSSION AND PRELIMINARY RESULTS

### Membrane Metallization

After the metallization, the Ti/Au bilayer films on the pnc-Si membrane were imaged to ensure that pores have not been totally occluded. TEM images of the membrane without metal and with 20 nm of evaporated titanium and gold are shown in Fig. 3.

The pnc-Si membrane in Fig.3a has a distribution of pore sizes. Deposition of the metal layer on this pnc-Si membrane (Fig. 3b) covers all the smaller pores and decreases the size of the larger pores as metal coats the edges of the pores. However, the membrane is still porous and permeable after the metal coating and can be used as a conductive plate for a capacitive sensor. The porosity drops from 13-14 % to 3-4 % for membranes with an average pore diameter of 28 nm and from 4 % to < 1 % for membranes with an average pore diameter of 14 nm as measured by gas permeability experiments.

**Figure 3**. a) TEM of pnc-Si membrane with low porosity; b) TEM image of the same membrane with 10 nm of titanium and 10 nm gold coatings.

### Optical Measurement of Vapor Permeation

The pore diameter in the pnc-Si membrane is approximately the same as the membrane thickness so the membrane overall exhibits negligible flow resistance. An experiment was conducted on the pnc-Si membrane/PDMS system to determine if the vapor above the pnc-Si membrane can permeate through pores, reach the polymer and induce changes in it. The goal was to measure the swelling of the PDMS induced by xylene vapor [10]. A structure consisting of a PDMS layer sandwiched between silicon and pnc-Si membrane chips was fabricated as the sensor structure described above in Fig. 2. It was then subjected to xylene vapor and the change in the surface height before and after the solvent introduction was measured.

Optical profilometry was used to measure the induced swelling of the Si/PDMS/pnc-Si structure. Optical profilometery is a white light interferometry technique. White light first passes through a beam splitter and is directed to the sample surface and a reference mirror. Reflected light from both surfaces is later recombined to produce interference fringe pattern which gives information about the surface contour of the sample. In this study a Veeco Wyko optical profiler was used.

**Figure 4.** 3D optical profilometry images of a pnc-Si/PDMS/Si structure taken from the well-side of the pnc-Si membrane: a) initial structure before exposure to vapor; b) immediately after exposure to xylene vapor; c) approximately 2 minutes after the vapor is removed.

Fig. 4 shows 3D optical profilometry images of a pnc-Si/PDMS/Si structure taken from the well-side of pnc-Si membrane. The square window area in these images is the pnc-Si membrane covering PDMS layer. The surface height is represented by the color scale. The wall-like area around the membrane window is an artifact of the imaging as the optical profiler picks up the signal reflected not only from the square membrane area but also from the walls of the silicon well which was etched to expose the freestanding pnc-Si membrane. The deflection of the structure was observed before, during and after exposure to xylene vapor. Due to xylene vapor permeation through the pores, PDMS swells and its thickness under the pnc-Si membrane increases. A 0.38 μm increase in thickness of the initial PDMS layer (Fig. 4a) under the membrane was measured when xylene vapor was introduced (Fig. 4b) into the system. After the vapor source was removed the membrane returned to its initial state as shown in Fig.4c.

### Capacitance Measurements of Vapor Permeation

To measure the capacitance of the complete sensor a computer was networked with a HP4275A multifrequency LCR meter controlled by Labview to collect real-time sensing data. All measurements were done in the parallel capacitive mode, with an AC signal of 10 kHz and voltage amplitude of 100 mV. The measurement setup is shown in Fig.5a. Conductive silver epoxy was used to make wire connections to the sensor electrodes. The sensor chip was connected to the LCR meter with the backside well facing up to ensure vapor permeation through the porous membrane. Real-time data was collected during solvent introduction. A pipette was used to dispense small amounts of liquid solvents into the container where it evaporated. An immediate change in capacitance was observed. The sensor was then taken out to allow the capacitance to return to the baseline value, and then exposed to the solvent vapor again. This procedure was repeated several times.

The capacitance between plates of a parallel-plate capacitor is defined as $C=\varepsilon_0\varepsilon_r A/d$, where $\varepsilon_0$ is the permittivity of vacuum, $\varepsilon_r$ is the dielectric constant of the dielectric layer, A is the overlap area between two plates, d is the separation between plates. The effects of polymer swelling (d+Δd) and change of dielectric constant ($\varepsilon_r+\Delta\varepsilon_r$) may cancel each other. To maximize the capacitive response the polymer and solvent were chosen so that one of these mechanisms is dominant. Test solvents, including hexane, toluene, and acetone, were selected for this experiment as they are readily available, quick to evaporate and induce only one of two described above changes in PDMS. For hexane and toluene, polymer swelling dominates whereas for acetone vapor, dielectric constant change dominates.

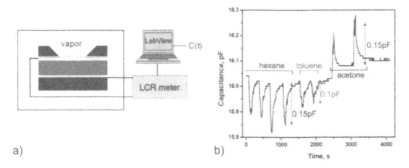

a)                                    b)

**Figure 5.** a) Capacitance measurement setup; b) real-time capacitance data with repetitive introduction of hexane, toluene and acetone vapor.

Preliminary experimental capacitance data is shown in Fig. 5b. First the sensor was repeatedly exposed to hexane and then toluene vapor of approximate concentrations of 800 ppm and 1000 ppm respectively. These solvents swell the PDMS, which increases plate separation and decreases the capacitance. Hexane swells PDMS more than toluene, which is known from comparing their solubility parameters $\delta$ ($\delta_{PDMS} = \delta_{hexane}$ and $\delta_{PDMS} < \delta_{toluene}$) [10]. The dielectric constant of these two solvents is close to the dielectric constant of PDMS and the net change in dielectric constant is negligible. The changes in capacitance induced by hexane and toluene were 0.15 pF and 0.1 pF accordingly. Next the sensor was exposed to approximately 1200 ppm of acetone vapor. The dielectric constant of acetone ($\varepsilon_{r\,acetone} > \varepsilon_{r\,PDMS}$) is much higher than that of PDMS resulting in an increase of total sensor capacitance after acetone exposure. PDMS swelling by acetone is negligible [10]. The increase was measured to be 0.15 pF. The differences in capacitance values for the same vapor exposures occur because vapor concentrations were not precisely calibrated in these initial experiments and longer recovery times are needed to ensure that all the vapor diffuses out of the PDMS matrix.

## CONCLUSIONS

A novel parallel-plate capacitive sensor that uses novel ultrathin pnc-Si membrane material covered with titanium and gold as an electrode was fabricated. The metal deposition is shown to keep the larger pores open and to allow the sensing analyte to permeate. The swelling caused by vapor permeation was measured by optical profilometry. Demonstration of real-time capacitance response of this sensor was presented for three solvents that induce two different mechanisms of capacitance changes, based on dielectric thickness or dielectric constant change. Future work will include fabrication of a flow-through capacitive sensor using two membranes as permeable plates. Since membrane and sensor fabrication is compatible with standard microfabrication processes there is significant potential for an entire detection system to ultimately be integrated on a single silicon chip.

## ACKNOWLEDGMENTS

This work was supported by the National Science Foundation (CBET 0707795 and ECCS 0707795). Device fabrication and characterization was performed at the University of Rochester and Rochester Institute of Technology Semiconductor Microfabrication Facility.

## REFERENCES

1. Striemer, C. C., Gaborski, T. R., McGrath, J. L. & Fauchet, P. M. *Nature* **445**, 749-753 (2007)
2. Li, J.R., Xu, J.R., Zhang, M.Z. *et al. Carbon* **41**, 2353 (2003)
3. Patel, S. V., Mlsna, T. E. *et al. Sens. Actuators* B **96**, 541–553 (2003)
4. Kitsara, M., Goustouridis, S. *et al. Sens. Actuators* B **127**, 186-192 (2007)
5. Kummer, A.M., Hierlemann, A., Baltes, H. *Analytical Chemistry* **76**, 2470-2477 (2004)
6. Hierlemann, A. *et al. Sens. Actuators* B **70**, 2 (2000)
7. McGill, R. A., *et al. Sens. Actuators* B **65**, 10 (2000)
8. Grate, J. W., Wise, B. M., Abraham, M. H. *Analytical Chemistry* **71**, 4544 (1999)
9. Patel, S.V. *et al Int. J. of P. and An. Chem.* **76**, 872-877 (2008)
10. Lee, J.N., Park, C., Whitesides, G. M. *Analytical Chemistry* **75**, 6544-6554 (2003)

Mater. Res. Soc. Symp. Proc. Vol. 1190 © 2009 Materials Research Society          1190-NN03-11

# Spiral Photonic Actuator

K.-Y. Jin,[1] S.-K. Park,[1] J.-H. Jang,[2] C. Y. Koh,[2] M. J. Graham,[3] S.-J. Park,[1] C. Nah,[1] M.-H. Lee,[1] S. Z. D. Cheng,[3] E. L. Thomas[2] , and K.-U. Jeong[1,2,3]
[1]Polymer Materials Fusion Research Center and Department of Polymer-Nano Science & Technology, Chonbuk National University, Jeonju, Jeonbuk 561-756, South Korea
[2]Institute for Soldier Nanotechnologies and Department of Materials Science and Engineering, Massachusetts Institute of Technology, Cambridge, MA, USA
[3]Department of Polymer Science, The University of Akron, Akron, OH, USA

## ABSTRACT

There has been a significant effort to create spiral sensors by changing either the periodic d-spacing of the structure or the dielectric constants of the materials by combining the multi-faceted environmental responsiveness of polymer hydrogels with dielectrical structures.[1] Reversible spiral switches with dimensional functionalities that respond to chemical environment were constructed. When the spiral photonic actuator was swollen in hydrophilic acetic acid, right-handed spiral structures are formed, while the spiral photonic actuator was swollen in hydrophobic hexane, left-handed spiral structures are formed. All actuators returned back to the transparent planar state after deswelling processes. These reversible spiral photonic actuators can be applied in the application of mechanical actuators, electrical devices, and optical components.

## INTRODUCTION

Soft polymeric materials have been widely used to fabricate intelligent soft materials and complex structures for various practical applications. Examples include photonic crystals,[2-6] actuators,[7-11] and spiral structures,[12-17] which are of particular interests for the next generation of biological and electro-optical technologies. The three-dimensional (3D) structure is applicable to many important geometric structures in biological polymers such as proteins and deoxyribonucleic acid. In addition to biology, spiral structures have been intensively studied and developed in electro-optical material science and technology.[18-20] The simplest bio-mimetic actuator is the 1D cantilever. Among many materials, polymers have been key building blocks in the fabrication of actuator because of their ability to change shape and size due to environmental changes such as ionic character, pH, temperature and solvent.[21-25] One of the most well known photonic crystals is the opal which has a periodic structure. The regular photonic crystals can affect the propagation of electromagnetic waves in the same way as the periodic potential in a semiconductor crystal affects the motion of electrons, by defining allowed and forbidden electronic energy bands.[26] The absence of allowed propagating modes for a range of wavelengths inside the structures gives rise to distinct optical phenomena such as inhibited spontaneous emission, and lossless reflection enabling low-loss-waveguiding.[27] Recently, there has been significant effort put into creating photonic actuators by changing either the periodic d-spacing of the structure or the dielectric constants of the materials.[28-30]

**EXPERIMENT**

Reversible spiral photonic actuators were fabricated, which can respond to both hydrophilic and hydrophobic environments by changing shape through a symmetry breaking and color through the periodicity of the photonic crystal. A schematic diagram of the fabricating a reversible spiral photonic actuator that undergoes simultaneously color and shape changes is shown in Fig. 1a. To fabricate spiral photonic actuators, colloidal silica spheres with diameters of 270 nm (Bangs Laboratories) were first assembled into thick near single crystalline opals with a FCC structure[31, 32] on a fluorinated silane layer treated mold. Scanning electron microscope (SEM) images of the assembled silica colloidal FCC structure along the [111] and [01$\bar{1}$] zones are shown in Fig. 1b and 1c, respectively. The reversible actuation processing is induced by the different swelling ratios of the two layers in the same solvent. A poly (dimethylsiloxane) (PDMS) and a polyurethane (PU)/2-hydroxyethylmethacrylate (HEMA) precursor were selected as components of the bilayer because of their flexibility, optical transparency, and dramatically different swelling responses to various solvents. The HEMA content in the PU/HEMA layer was optimized at 30 wt%. The next step was to infiltrate the silica colloid with the refractive index matched hydrophobic PDMS precursor and then to cure it at 65 °C for 12 h. The corresponding SEM [111] zone image of the PDMS infiltrated silica colloid is shown in Fig. 1d. Finally the hydrophilic layer is applied to the hydrophobic PDMS/silica photonic crystal to fabricate the bilayer structure. Strong adhesion between the PDMS and PU70/HEMA30 layers was achieved by creating a methacryloxypropyltrimethoxysilane monolayer on the PDMS by the monolayer self-assembly on the oxygen plasma treated PDMS surface and then applying the hydrophilic PU70/HEMA30 layer. The methacryloxypropyltrimethoxysilane monolayer serves as a strong adhesion between the two layers. Finally parafilm and glass plate were placed on top of the mold during the UV-curing process of the hydrophilic PU70/HEMA30 layer in order to ensure a constant sample thickness. A photograph of the fabricated spiral photonic actuator in the dry state is shown in Fig. 1e.

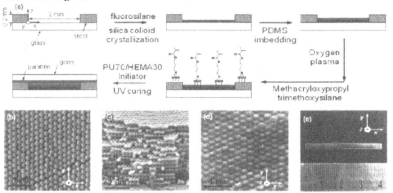

**Fig. 1** (a) Schematic diagrams of the fabricated processes for the spiral photonic actuator. Scanning electron microscope (SEM) images of silica colloidal/air photonic crystal on (b) [111] and (c) [01$\bar{1}$] zones, respectably, and (d) of the PDMS-imbedded silica photonic crystal on [111] zone. (e) A photograph of the spiral photonic actuator (Reproduced by permission of The Royal Society of Chemistry).[1]

In order to evaluate the photonic stop-bands of the photonic actuator, reflectivity was measured utilizing an optical microscope equipped with a fiber-optic spectrometer and a near infrared spectrometer using a silver-coated metallic mirror as a 100% reference. Fig. 2 shows experimental reflectivity spectra obtained from the [111] zone of a photonic actuator in different chemical environments. The 270 nm silica colloidal photonic crystals showed 60% reflectivity at a wavelength of 590 nm. Because the refractive index of the silica colloid and the PDMS are matched (n ≅ 1.43), there is no reflection peak and the sample appears transparent after infiltration of the PDMS into the silica photonic crystal (Fig. 1e and 2b).

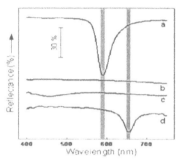

**Fig. 2** Experimental reflectivity spectra of the fabricated photonic crystals: (a) silica, (b) PDMS/silica, (c) PDMS/silica in hexane, and (d) PDMS/silica in acetic acid (Reproduced by permission of The Royal Society of Chemistry).[1]

The swelling of the PDMS in the proper solvent leads to volume change in the photonic crystal and thus a red shift in the transmission notch. This swollen silica colloid/PDMS photonic crystal exhibited a reflection peak with its maximum intensity (20%) centered at 659 nm in acetic acid and reflected a broad, low intensity peak (7%) at 459 nm in hexane (Fig. 2). It should be noted that the reason for the weak transmission notches stems from the polycrystallinity of the opal in the silica-PDMS layer as experimentally observed upon infiltration in addition to defects in the colloidal template. Using averaged swelling data (Fig. 3), we calculate the reflection peaks and d-spacing of the (110), (200), and (120) planes: 801 nm (d = 277 nm), 567 nm (d = 196 nm) and 507 nm (d = 175 nm), respectively. These results match the sky-blue (Fig. 4a), yellow-green (Fig. 4b), and pink (Fig. 4c) colors seen at different viewing angles when the samples are back-lit in acetic acid. The functionality and sensitivity of the spiral photonic actuator was confirmed by exposing the bilayer composite to hydrophilic acetic acid (Fig. 4a-4c) and hydrophobic hexane (Fig. 4d-4f). Kinetic swelling data for different solvents with respect to time is shown in Fig. 3. The maximum swelling ratio is defined as $V_{max}/V_0$, where $V_{max}$ is the maximum expanded volume at swelling time, $t = t_{max}$ and $V_0$, is the initial volume at $t = t_0$. When the switch is swollen in the hydrophilic acetic acid solvent, $V_{max}/V_0 = 1.10$ for hydrophobic PDMS layer and $V_{max}/V_0 = 2.05$ for hydrophilic PU70/HEMA30. The asymmetric volume change results in a shape change and the swelling of the PDMS leads to an expansion of the photonic crystal and a refractive index change resulting in a color change. The spiral wrapped around itself 3 to 4 times at this swelling ratio and switch length. The increased refractive index contrast enabled the structure to reflect light with the wavelength of the notch depend on the optical thickness of the expanded lattice.

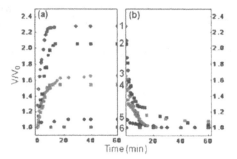

**Fig. 3** The expanded volume ratio ($V/V_0$) with respect to (a) swelling and (b) deswelling times for (1) PDMS in hexane, (2) PU70/HEMA30 in acetic acid, (3) PDMS in ethyl acetate, (4) PU70/HEMA30 in ethyl acetate, (5) PDMS in acetic acid, and (6) PU70/HEMA30 in hexane (Reproduced by permission of The Royal Society of Chemistry).[1]

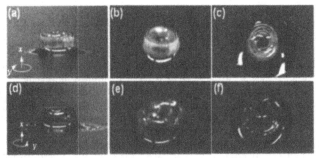

**Fig. 4** A right-handed spiral photonic actuator in acetic acid: (a) side view, (b) 45°-tilted view, and (c) top view, respectively. A left-handed spiral photonic actuator in hexane: (d) side view, (e) 45°-tilted view, and (f) top view, respectively. The geometric description of the sample is identical to that of Fig. 1a (Reproduced by permission of The Royal Society of Chemistry).[1]

## DISCUSSION

In acetic acid, a red right-handed spiral is formed as shown in Fig. 4a-4c. The silica colloid/PDMS photonic crystal actuator returned to a transparent planar structure when the acetic acid completely evaporated. When the transparent photonic crystal actuator is swollen in hydrophobic hexane, a bluish left-handed spiral is formed as shown in Fig. 4d-4f. The $V_{max}/V_0$ of the hydrophobic PDMS layer in hydrophobic hexane is 2.28, while the $V_{max}/V_0$ of hydrophilic PU70/HEMA30 in hexane is 1.00. In this case, the photonic crystal structure in PDMS is also expanded resulting in a loss of long-range spatial correlation between the silica spheres thereby suppressing Bragg diffraction. Therefore, the blue color is caused by the scattering of uncorrelated dielectric particles. This result explains the small broad reflectivity peak and the independence of color on viewing angle (Fig. 4d-4f). When the hexane is completely evaporated, the silica colloid/PDMS actuator becomes a transparent planar structure. This suggests that the loss of correlation is due to the swelling gradient which is not random, local, and irreversible swelling. This indicates that the color functionality of the photonic crystal switch can operate by two distinct reversible mechanisms depending on the swelling differential and the location of the opal in the bilayer. In addition this color-tunable photonic actuator is fully reversible up to more than 5 times without any optical and mechanical property changes.

In order to demonstrate the functionality and sensitivity of the switch to hydrophobicity, ethyl acetate was selected as a solvent. The $V_{max}/V_0$ of the hydrophobic PDMS layer in ethyl acetate is 1.65, and the $V_{max}/V_0$ of hydrophilic PU70/HEMA30 in ethyl acetate is 1.55. In this case the PDMS layer is significantly swollen but the small swelling differential translates into a smaller curvature. The result is a left-handed spiral structure where the reflected colors change with viewing angle like in acetic acid. In addition, kinetic swelling experiments as shown in Fig. 3 indicate that each solvent can completely swell the polymers within minutes. This is adequately fast for the desired switching speeds needed for practical application. Faster speeds could be achieved by using materials or solvents with a greater swelling differential or thinner samples.

## CONCLUSIONS

Reversibly color-tunable spiral photonic actuators with both dimensional and optical functionalities that respond to chemical environment were constructed. When the transparent photonic actuator is swollen in hydrophilic acetic acid, right-handed spiral actuators that exhibit angularly dependent colors from Bragg diffraction are formed. However, when the transparent photonic actuator is swollen in hydrophobic hexane solvent, a left-handed spiral actuator with an angularly independent bluish color is formed. After deswelling, the spiral photonic actuators returned back to the transparent planar state. This reversible color-tunable spiral photonic actuator can be useful as mechanical actuators, electrical devices, and optical components.

## ACKNOWLEDGMENTS

This work was mainly supported by KRF-2007-331-D00119 and Fundamental R&D Program for Core Technology of Materials funded by the Ministry of Knowledge Economy, Korea and also partially supported by ISN (W911NF-07-D-0004), NSF CMS-0556211, DMR-0804449 and DMR-0516602. More detail description of this work can be found in Journal of Material Chemistry, 2009, 19, 1956-1959.

## REFERENCES

1. K.-U. Jeong, J.-H. Jang, C. Y. Koh, M. J. Graham, K.-Y. Jin, S.-J. Park, C. Nah, M.-H. Lee, S. Z. D. Cheng, and E. L. Thomas, J. Mater. Chem., **19**, 1956-1958, (2009).
2. P. Vukusic, J. R. Sambles and C. R. Lawrence, Nature, **404**, 457, (2000).
3. M. Maldovan and E. L. Thomas, Nature Mater., 3, 593, (2004).
4. N. V. Dziomkina and G. J. Vancso, Soft Matter, 1, 265, (2005).

5. A. Yethiraj, Soft Matter, **3**, 1099, (2007).
6. I. Musevic and M. Skarabot, Soft Matter, **4**, 195, (2008).
7. S. V. Ahir and E. M. Terentjev, Nature Mater., **4**, 491, (2005).
8. V. H. Ebron, J. W. Yang, D. J. Seyer, M. E. Kozlov, J. Y. Oh, H. Xie, J. Razal, L. J. Hall, J. P. Ferraris, A. G. MacDiarmid and R. H. Baughman, Science, **311**, 1580, (2006).
9. R. D. Rey, Soft Matter, **3**, 1349, (2007).
10. S. Haider, S. Y. Park and S. H. Lee, Soft Matter, **4**, 485, (2008).
11. P. D. Thornton, R. J. Mart, S. J. Webb and R. V. Ulijn, Soft Matter, **4**, 821, (2008).
12. S. F. Mason, Nature, **311**, 19, (1984).
13. C. Y. Li, S. Z. D. Cheng, J. J. Ge, F. Bai, J. Z. Zhang, I. K. Mann, F. W. Harris, L.-C. Chien, D. Yan, T. He and B. Lotz, Phys. Rev. Lett., **83**, 4558, (1999); J. Am. Chem. Soc., **123**, 2462, (2001).
14. H. Shen, K.-U. Jeong, H. Xiong, M. J. Graham, S. Leng, J. X. Zheng, H. Huang, M. Guo, F. W. Harris and S. Z. D. Cheng, Soft Matter, **2**, 232, (2006).
15. R. J. Mart, R. D. Osborne, M. M. Stevens and R. V. Ulijn, Soft Matter, 2006, 2, 822;
16. J. L. Gornall and .E M. Terentjev, Soft Matter, **4**, 544, (2008).
17. G. Zubay, *Biochemistry*, 2nd ed., (Macmillan Publishing Company, New York, 1988), pp. 845–1151.
18. R. Purrello, Nature Mater., **2**, 216, (2003).
19. K.-U. Jeong, S. Jin, J. J. Ge, B. S. Knapp, M. J. Graham, J. Ruan, M. Guo, H. Xiong, F. W. Harris and S. Z. D. Cheng, Chem. Mater., **17**, 2852, (2005); Macromolecules, **38**, 8333, (2005); Polymer, **47**, 3351, (2006); Adv. Mater., **18**, 3229, (2006); Chem. Mater., **18**, 680, (2006).
20. D.-K. Yang, K.-U. Jeong and S. Z. D. Cheng, J. Phys. Chem. B, **112**, 1358, (2008).
21. R. D. Kamien, Science, **315**, 1083, (2007).
22. Y. Klein, E. Efrati and E. Sharon, Science, **315**, 1116, (2007).
23. Y. Osada, H. Okuzaki and H. Hori, Nature, **355**, 242, (1992).
24. E. Smela, Adv. Mater., **15**, 481, (2003).
25. S. J. Kim, G. M. Spinks, S. Prosser, P. G. Whitten, G. G. Wallace and S. I. Kim, Nature Mater., **5**, 48, (2006).
26. J. Y. Huang, X. D. Wang and Z. L. Wang, Nano Lett., **6**, 2325, (2006).
27. J. D. Joannopoulos, R. D. Meade, J. N. Winn, *Photonic Crystals: Molding the Flow of Light*, (Princeton University, 1995).
28. J. H. Holtz and S. A. Asher, Nature, **389**, 829, (1997).
29. J. M. Weissman, H. B. Sunkara, A. S. Tse and S. A. Asher, Science, **274**, 959, (1996).
30. M. Fialkowski, A. Bitner and B. A. Grzybowski, Nature Mater., **4**, 93, (2005).
31. P. Jiang, J. F. Bertone, K. S. Hwang and V. L. Colvin, Chem. Mater., **11**, 2132, (1999).
32. P. Jiang, K. S. Hwang, D. M. Mittleman, J. F. Bertone and V. L. Colvin, J. Am. Chem. Soc., **121**, 11630, (1999).

Mater. Res. Soc. Symp. Proc. Vol. 1190 © 2009 Materials Research Society 1190-NN03-20

# Rational Design of Nanostructured Hybrid Materials for Photovoltaics

Ioan Botiz and Seth B. Darling
Center for Nanoscale Materials, Argonne National Laboratory, 9700 S. Cass Ave.,
Argonne, IL 60439, U.S.A.

## ABSTRACT

To develop efficient organic and/or hybrid organic-inorganic solar energy devices, it is necessary to use, among other components, an active donor–acceptor layer with highly ordered nanoscale morphology. In an idealized morphology, the effectiveness of internal processes is optimized leading to an efficient conversion of photons to electricity. Using a poly(3-hexylthiophene)-*block*-poly(L-lactide) rod-coil block copolymer as a structure-directing agent, we have rationally designed and developed an ordered nanoscale morphology consisting of self-assembled poly(3-hexylthiophene) donor domains of molecular dimension, each of them separated by fullerene $C_{60}$ hydroxide acceptor domains. Using this morphological control, one can begin to probe structure-property relationships with unprecedented detail with the ultimate goal of maximizing the performance of future organic/hybrid photovoltaic devices.

## INTRODUCTION

Traditional photovoltaic materials are well-suited to some markets, these technologies are currently too expensive to implement on a global scale, so scientists are directing fresh attention toward new generations of organic and/or hybrid low-cost photovoltaic (PV) devices.[1,2] Demonstrated efficiencies of such PV energy devices are far short of the thermodynamic limit. Efficient conversion of photons to electricity in organic and hybrid materials depends on optimization of factors including light absorption,[3] exciton separation,[4] and charge carrier migration.[5] Various PV device concepts have been developed, including polymer-based solar cells. The most efficient such devices are made by blending a semiconducting polymer either with fullerenes,[2,6] other polymers,[7] or nanocrystals.[8] One conclusion that appears when dealing with polymer-based PV devices is that donor–acceptor bulk heterojunction (BHJ) devices address key internal processes, but disorder on the nanoscale results in inefficiencies due to exciton recombination and poor mobility. In order to develop high-performance organic and/or hybrid organic-inorganic solar energy devices, it is necessary to control the active layer morphology on the nanoscale.[9]

One promising approach is to design block copolymers possessing one or more optoelectronic active blocks.[10] These copolymers can undergo a microphase separation process that leads to ordered morphologies on surfaces[11] characterized by nanostructured domains of characteristic distance tuned by the polymer block lengths. The goal of the work presented here is to experimentally implement the idea of morphological control using common PV active component materials: poly(3-hexylthiophene) and $C_{60}$. While the standard approach is to mix these two compounds and manipulate processing conditions to try to optimize nanoscale phase separation in the BHJ layer, we have used a novel fabrication process that relies on self-organization to generate idealized morphology. The starting material is a poly(3-hexylthiophene)-*block*-poly(L-lactide) (P3HT-*b*-PLLA) linear rod-coil diblock copolymer (Figure 1a), which has been recently reported by Hillmyer et al.[12] The molecular weights of the two blocks were selected to encourage self-assembly into a lamellar structure. P3HT is a semiconducting polymer, whereas the role of the PLLA is described below. The P3HT-*b*-PLLA is designed to be used both as an optoelectronically active material and as a structure-directing agent to pattern active material into ordered nanostructures (schematically depicted in Figure 1b). The characteristic periodicity, which will eventually translate into the donor–acceptor length scale, for P3HT-*b*-PLLA was designed to be approximately 15 nm, i.e. comparable to the exciton diffusion length throughout the active layer.[13] This periodicity can be readily tuned by adjusting the overall molecular weight.

An essential advantage of using P3HT-*b*-PLLA is the fact that the coiled PLLA block is biodegradable.[14] Therefore, once the microphase separation process is complete and the ordered nanoscale morphology is obtained, the PLLA block can be readily removed using the procedure described in the Experimental section. The removal of PLLA opens new possibilities. For example, we can use the former PLLA domains as vessels for an acceptor material like fullerene hydroxide ($C_{60}$). The more traditional form of $C_{60}$, [6,6]-phenyl-$C_{61}$-butyric acid methyl ester (PCBM), is not suitable in this process because organic solvents would destroy the nanostructured P3HT template. This morphology, consisting of alternating donor–acceptor nanostructured domains, is expected to enhance the effectiveness of internal processes compared to a less ordered BHJ morphology obtained when simply mixing P3HT homopolymer with PCBM. The fabrication time is comparatively longer than traditional BHJ approaches due to the long etching and electron-acceptor infiltration steps, but the goal here is rational morphological control for structure-property investigations that cannot be achieved using traditional methods.

## EXPERIMENT

The P3HT-*b*-PLLA linear diblock copolymer was purchased as a custom synthesis from Polymer Source, Inc. and used without further purification. The number average molecular weight was $M_n=3300$ for P3HT and $M_n=4100$ for PLLA. The polydispersity index ($M_w/M_n$, PDI) was 1.28, as measured by size exclusion chromatography. Fullerene $C_{60}$ hydroxide [chemical structure: $C_{60}(OH)_{22-26}$] was purchased from Materials Technologies Research, Ltd. This acceptor material is water soluble at pH≥8. Thin solid films with an average thickness of ~70 nm (measured by ellipsometry and profilometry) were obtained by spin-casting chloroform polymer solution onto ITO-covered glass and $SiO_2$. The surface of the thin films was characterized in detail by tapping-mode atomic force microscopy [Nanoscope V, Veeco, USA]. X-ray diffraction data (XRD) were recorded using a Bruker D8 Discover analytical x-ray system. Attenuated total reflectance (ATR) Fourier-transform infrared (FT-IR) spectra were recorded using a Vertex 70 spectrometer from Bruker and a 20× ATR objective (Ge-crystal). The etching process used to selectively remove the biodegradable PLLA block consisted of treatment of thin polymer films in 0.5 M NaOH solution (60:40 v/v water/methanol). Thin films were submerged in NaOH solution for 5 days and then carefully rinsed with water in order to completely remove degraded PLLA fragments and Na. After rinsing, thin films were dried overnight at room temperature.

## RESULTS and DISCUSSION

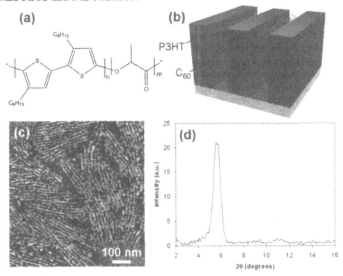

**Figure 1.** (a) Chemical structure of P3HT-*b*-PLLA linear rod-coil diblock copolymer; (b) Schematic of ordered nanoscale morphology consisting self-assembled P3HT donor domains (lamellae), each separated by $C_{60}$ acceptor domains; (c) TM-AFM phase image showing an ordered morphology obtained by spin-casting of P3HT-*b*-PLLA on ITO-covered glass (polymer film thickness of 69±3 nm). AFM image was taken after annealing film in chloroform vapor for 90 minutes; (d) XRD pattern of a thin film of P3HT-*b*-PLLA spin-cast from chloroform solution on $SiO_2$ substrate.

The atomic force microscopy (AFM) phase image depicted in Figure 1c shows the resulting morphology obtained after spin-casting P3HT-*b*-PLLA chloroform solution on ITO-covered glass. We observe a morphology comprised of periodic parallel stripe domains, as designed. Domains of lighter color correspond to the P3HT block while the darker domains are the PLLA amorphous block, suggesting a lamellar structure with the alternating domains oriented perpendicular to the substrate. Analyzing the AFM image of Figure 1c in more details, we have found (not shown) a characteristic distance of 16±3 nm which corresponds well with the calculated molecular length scale of about 15 nm.

For solar cell applications, in which electrodes are situated on the substrate and on top of the active layer, respectively, order along the vertical direction is essential for efficient charge carrier migration whereas order in the plane of the film is less important. There are two reasons to believe the perpendicular morphology (observed at the film surface by AFM) extends down from the free interface to the substrate. The first is that a number of groups have completed rigorous structural studies on related rod-coil block copolymers in thin films[15] and they have shown the formation of continuous perpendicular lamellae (closely resembling those seen in Figure 1c) extending down from the free interface to the substrate. In addition to this collection

213

of results in the literature, there is a second reason to conclude that the perpendicular morphology extends throughout the film. The alternative possibility is that we have perpendicular lamellae at the free interface and parallel lamellae underneath. If such an arrangement did exist in P3HT-*b*-PLLA, the film would not survive the NaOH treatment that removes the PLLA domains. Removal of PLLA domains oriented parallel to the substrate would delaminate the film (which is not the case, i.e. most areas are populated by the perpendicular morphology). Further evidence for this morphology comes from diffraction data. Figure 1d depicts an XRD pattern of a P3HT-*b*-PLLA thin film spin-cast on a $SiO_2$ substrate. This pattern shows a strong peak at an angle $2\theta=5.6°$ corresponding to a lattice spacing of ~16 Å. These data indicate not only that the P3HT is crystalline but also that the P3HT polymer chains are oriented with their long axes parallel to the substrate, with that spacing representative of the periodicity between vertically stacked chains. Because of the immiscibility of the two constituent polymer blocks, it is unlikely that the PLLA chains would be oriented with their long axis otherwise than parallel on the substrate as well.

Once the ordered nanoscale morphology is obtained, there is no need for the PLLA block, which is used to simply direct structure formation by microphase separation, to remain. Therefore, the PLLA block was removed by employing a simple etching process. We have introduced two nanostructured films (69±3 nm thickness) in 0.5 M NaOH solution for 2 hours and 5 days, respectively. ATR FT-IR spectra were recorded before (after just 2 hours of etching) and after the complete etching of the PLLA block. The results are shown in Figure 2a. Here, one can clearly observe that several peaks (around 1050, 1095, 1135, 1195, 1270, 1450 and 1765 cm$^{-1}$, respectively) completely disappeared after 5 days of etching. According to the literature,[14] these peaks were assigned to the different stretching vibrations of PLLA block components. Moreover, the final spectrum matches well with that of pure P3HT. These results demonstrated the complete and successful removal of the PLLA block from thin structured films.

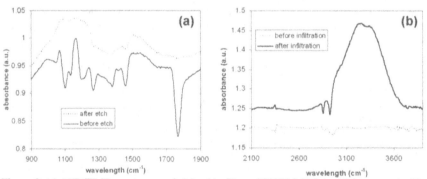

**Figure 2.** (a) ATR FT-IR spectra recorded for thin films of P3HT-*b*-PLLA which were etched in 0.5 M NaOH solution for 72 (top) and 2 hours (bottom), respectively; (b) ATR FT-IR spectra recorded for a thin film of P3HT-*b*-PLLA (spectrum on bottom) and for a film of P3HT-*b*-PLLA which was etched for 5 days in 0.5 M NaOH solution and then dip-coated in aqueous solution of C60 (15 g/L, pH=8.7) for 72 hours (spectrum on top).

Once the PLLA block has been removed, one can fill the empty spaces with electron acceptor material, such as $C_{60}$. This procedure has been performed by dip-coating the thin film of nanoporous P3HT in an aqueous solution of the $C_{60}$ for 72 hours. Again, we have recorded the ATR FT-IR spectra before and after the dip-coating process. Results are summarized in Figure 2b. Here, one can observe that an intense peak at about 3360 cm$^{-1}$ appeared after 72 hours of dip-coating of the thin film in aqueous solution of $C_{60}$. According to the FT-IR spectrum provided by the vendor (not shown), this peak is assigned to stretching vibrations of the $C_{60}$ acceptor material.

The above data demonstrate that $C_{60}$ material was transferred to the nanostructured thin films of P3HT. Because polar surface moieties of the $C_{60}$ acceptor material do not possess an affinity for essentially non-polar P3HT, it is likely that this material is located within the empty domains created after the removal of PLLA (as opposed to on top of the P3HT stripes). Ongoing experiments will focus on fabrication of prototype solar energy devices and will establish the dip-coating time needed to transfer the optimum amount of $C_{60}$ acceptor material.

## CONCLUSIONS

A rod-coil block copolymer composed of an electron-donating block and a sacrificial inactive block has been used to template an ordered nanoscale material system with potential use in next-generation photovoltaics. Upon removal of the lactic acid-based sacrificial block, the remaining nanoscale voids can be infiltrated with electron accepting materials. By controlling the nanoscale morphology using block copolymer self-assembly, it is possible to engineer an idealized structure in which exciton separation and charge migration can be optimized without sacrificing light absorption. Future work will incorporate this idealized active layer into prototype devices for comparison with traditional, disordered bulk heterojunctions.

## ACKNOWLEDGMENTS

Use of the Center for Nanoscale Materials was supported by the U.S. Department of Energy, Office of Science, Office of Basic Energy Sciences, under Contract No. DE-AC02-06CH11357.

## REFERENCES

1    B. C. Thompson and J. M. J. Fréchet, Angew. Chem. Int. Ed. **47**, 58 (2008); S. E. Shaheen, C. J. Brabec, N. S. Sariciftci et al., Appl. Phys. Lett. **78**, 841 (2001); W.U. Huynh, J.J. Dittmer, and A.P. Alivisatos, Science **295**, 2425 (2002); P. Peumans, S. Uchida, and S. R. Forrest, Nature **425**, 158 (2003); G. Li, V. Shrotriya, Y. Yao et al., J. Appl. Phys. **98**, 043704 (2005); T. Shiga, K. Takechi, and T. Motohiro, Sol. Energy Mat. Sol. Cells **90**, 1849 (2006).

2    Y. Kim, S. Cook, S. M. Tuladhar et al., Nat. Mater. **5**, 197 (2006).

3    J. Dai, X. Jiang, H. Wang et al., Appl. Phys. Lett. **91**, 253503 (2007); A. Gadisa, W. Mammo, L. M. Andersson et al., Adv. Funct. Mater. **17**, 3836 (2007).

4    S. M. Lindner, S. Hüttner, A. Chiche et al., Angew. Chem. Int. Ed. **45**, 3364 (2006); C. J. Brabec, N. S. Sariciftci, and J. C. Hummelen, Adv. Funct. Mater. **11**, 15 (2001); J.S. Salafsky, W.H. Lubberhuizen, and R.E.I. Schropp, Chem. Phys. Lett. **290**, 297 (1998).

5  M. M. Mandoc, F. B. Kooistra, J. C. Hummelen et al., Appl. Phys. Lett. **91**, 263505 (2007).

6  M. Reyes-Reyes, K. Kim, and D. L. Carroll, Appl. Phys. Lett. **87**, 083506 (2005); F. Padinger, R. S. Rittberger, and N. S. Sariciftci, Adv. Funct. Mater. **13**, 85 (2003); G. Yu, J. Gao, J. C. Hummelen et al., Science **270**, 1789 (1995); W. Ma, C. Yang, X. Gong et al., Adv. Funct. Mater. **15**, 1617 (2005).

7  M. M. Koetse, J. Sweelssen, K. T. Hoekerd et al., Appl. Phys. Lett. **88**, 083504 (2006); J. J. M. Halls, C. A. Walsh, N. C. Greenham et al., Nature **376**, 498 (1995); M. M. Alam and S. A. Jenekhe, Chem. Mater. **16**, 4647 (2004).

8  S. A. McDonald, G. Konstantatos, S. Zhang et al., Nat. Mater. **4**, 138 (2005); N. C. Greenham, X. Peng, and A. P. Alivisatos, Phys. Rev. B **54**, 17628 (1996); C. Y. Kwong, A. B. Djurisic, P. C. Chui et al., Chem. Phys. Lett. **384**, 372 (2004).

9  X. Yang and J. Loos, Macromolecules **40** (5), 1353 (2007); S. Gunes, H. Neugebauer, and N.S. Sariciftci, Chem. Rev. **107** (4), 1324 (2007); J. K. J. van Duren, X. Yang, J. Loos et al., Adv. Funct. Mater. **14**, 425 (2004); S.-S. Sun, Sol. Energy Mater. Sol. Cells **79**, 257 (2003); S.B. Darling, J. Phys. Chem. B **112**, 8891 (2008).

10  M. Sommer, S. M. Lindner, and M. Thelakkat, Adv. Funct. Mater. **17**, 1493 (2007); X. L. Chen and S. A. Jenekhe, Macromolecules **29**, 6189 (1996); G. Tu, H. Li, M. Forster et al., Macromolecules **39**, 4327 (2006).

11  S.B. Darling, Prog. Polym. Sci. **32**, 1152 (2007).

12  B. W. Boudouris, C. D. Frisbie, and M. A. Hillmyer, Macromolecules **41**, 67 (2008).

13  J. J. M. Halls, K. Pichler, R. H. Friend et al., Appl. Phys. Lett. **68**, 3120 (1996).

14  P. N. Thanki, E. Dellacherie, and J-L. Six, Appl. Surf. Sci. **253**, 2758 (2006).

15  C.-A. Dai, W.-C. Yen, Y.-H. Lee et al., J. Am. Chem. Soc. **129**, 11036 (2007); T. Heiser, G. Adamopoulos, M. Brinkmann et al., Thin Solid Films **511-512**, 219 (2006); B.D. Olsen, D. Alcazar, V. Krikorian et al., Macromolecules **41**, 58 (2008).

Printed in the United States
By Bookmasters